Automating Quality Systems

Automating Quality Systems

A guide to the design and implementation of automated quality systems in manufacturing

J.D.T. Tannock

Lecturer in Manufacturing Systems, Department of Mechanical Engineering,
University of Bristol, UK

SPRINGER-SCIENCE+BUSINESS MEDIA, B.V.

First edition 1992

Originally published by Chapman & Hall in 1992

ISBN 978-94-010-5044-9 ISBN 978-94-011-2366-2 (eBook)
DOI 10.1007/978-94-011-2366-2

A catalogue record for this book is available from the British Library

Library of Congress Cataloging-in-Publication data available

Contents

Acknowledgements

This book arose from practical research work funded by the Application of Computers in Manufacturing Engineering (ACME) Directorate of the Science and Engineering Research Council. The research was undertaken in collaboration with two manufacturing companies, Du Pont Electronics and Rolls-Royce plc. Thanks go to all those in these companies, especially Andy Phillips at Du Pont and Nick Orchard at Rolls-Royce, who have helped make the work possible.

I must also thank all at Bristol Polytechnic and the University of Bristol who have been involved with the research, especially George Trmal and Chris Earl for their guidance and help. I must also acknowledge the efforts of Ralph Wort and Barbara Savage, researchers on the Du Pont project, and David Cox and Huaming Lee, on the Rolls-Royce project.

Finally I must thank all those others who have welcomed me into their companies to examine quality assurance practice and automated inspection systems.

Preface

Quality is a topical issue in manufacturing. Competitive quality performance still eludes many manufacturers in the traditional industrialized countries. A lack of quality competitiveness is one of the root causes of the relative industrial decline and consequent trade imbalances which plague some Western economies.

Many explanations are advanced for poor quality performance. Inadequate levels of investment in advanced technology, together with insufficient education and training of the workforce, are perhaps the most prominent. Some believe these problems are caused by a lack of awareness and commitment from top management, while others point to differences between industrial cultures.

The established remedy is known as Total Quality Management (TQM). TQM requires a corporate culture change, driven from the top, and involving every employee in a process of never-ending quality improvement aimed at internal as well as external customers. The techniques deployed to achieve TQM include measures to improve motivation, training in problem-solving and statistical process control (SPC).

Quality is, however, only one of the competitive pressures placed upon the manufacturer by the modern global economy. It is also imperative to remain economical and efficient, while increasing the flexibility and responsiveness of the design and manufacturing functions. Here the reduction or elimination of stock is of great importance, particularly as financial interest rates in the less successful manufacturing nations are frequently high. Product life cycles must become ever more compressed in response to the phenomenal design-to-manufacture performance of some Pacific rim economies.

In manufacturing, two current trends are widely viewed as solutions to these challenges. One is flexible automation, intended to reduce the impact of high labour costs while enhancing responsiveness. The other is the just-in-time (JIT) manufacturing approach, targeted at the elimination of stocks, waste and waiting time.

JIT approaches tend to rely on simple people-orientated techniques such as *kanban*, which are in good accord with a TQM process. In practice, however, for many companies the complexity involved in the successful operation of an automated flexible manufacturing plant has required the wide use of computers, and now demands their effective integration. Computer-integrated manufacture (CIM) is the theme of development in this direction.

CIM thinking tends to concentrate on systems design and development of the hardware and software necessary to control and integrate automated functions. The human element is rarely considered. In contrast, a total quality (TQ) approach is founded on consideration of human factors. Perhaps as a result, automation in quality function activities is in its infancy compared with other areas of manufacturing.

Books on quality tend to fall on an axis between two extremes. At one end are comprehensive and voluminous textbooks replete with specialist detail concerning practical quality management issues. At the opposite extreme are TQ exhortations advocating a complete change in the culture of the business, making quality the responsibility of all rather than of the specialists. The literature reflects the 'split personality' of the quality field.

This book is not a textbook on quality assurance theory and practice, neither is it proposing anything new in TQM. Instead it attempts to develop another direction in quality thinking, towards the flexible automation and integration of quality function activities themselves. In this book I hope to present a framework for the automation of the quality function, in a manner compatible with the themes of TQM, and also with developments towards CIM.

The book is divided into four parts. In Part One, a brief survey is made of current themes in quality philosophy and strategy, and the case for automation is presented with particular reference to the implications of automated quality systems for CIM. Current methods of assessment of manufacturing quality performance are also examined, and an alternative approach is proposed, which is particularly suitable for automation.

In Part Two, the design and improvement of quality systems is explored. Different 'systems' approaches are contrasted, and a suitable technique for systems design proposed. A model information technology (IT) strategy for quality systems automation is suggested.

Part Three reviews the technology available for automated inspection. Coordinate measuring machines, machine vision systems, automatic test equipment and dimensional gauging are surveyed, with descriptions of typical applications of these technologies in manufacturing. In-process quality control is also outlined, with examples from machining and plastics manufacture, and a detailed case study describing an assembly force monitoring system is provided.

Part Four looks at quality data analysis and management, with an examination of SPC software, and the growing trend towards shop-floor data collection (SFDC) of quality data. The concept of the quality database is then examined, with a practical guide to its possible contents, and an approach to fully automated operation.

J. D. T. T.
University of Bristol

Illustrations

Acronyms

A/D	analogue to digital
AI	artificial intelligence
AMT	advanced manufacturing technology
ASIC	application-specific integrated circuit
ATE	automatic test equipment
BS	British Standard
BCD	binary coded decimal
CAD	computer-aided design
CADCAM	computer-aided design and manufacture
CAI	computer-aided inspection
CAM	computer-aided manufacture
CAPM	computer-aided production management
CAPP	computer-aided process planning
CAQ	computer-aided quality
CAQP	computer-aided quality planning
CASE	computer-aided software engineering
CCD	charge coupled device (machine vision camera)
CID	charge injection device (machine vision camera)
CIT	continuous improvement team
CIM	computer-integrated manufacture
CMM	coordinate measuring machine
CNC	computer numerical control (of machine tools)
COQ	cost of quality
CQDB	central quality database
DBMS	Database Management System
DCC	direct computer control (of CMM)
DMIS	Dimensional Measuring Interface Specification
EPROM	electronically programmable read only memory
FEO	for exposition only (of IDEF0 diagram)
FMS	flexible manufacturing system
4GL	fourth generation (computer) language

IC	integrated circuit
ICAM	integrated computer-Aided Manufacturing Program
IDEF	ICAM Definition Method
IDEF0	ICAM Definition Method Zero
I/O	input/iutput
IQCS	integrated quality control systems
IQS	integrated quality system
ISO	International Organization for Standardization
IT	information technology
JIT	just-in-time (manufacture)
LIMS	laboratory information management systems
LAN	local area network
LVDT	linear variable differential transformer (a dimensional transducer)
MS-DOS	Microsoft Disk Operating System
MRP	Materials Requirements Planning
MRPII	Manufacturing Requirements Planning
NDT	non-destructive testing
OCR	optical character recognition
OPT	optimized production technology
OR	operational research
ORACLE	Proprietary RDBMS
OSI	Open Systems Interconnection
PC	personal computer
PCB	printed circuit board
QA	quality assurance
QC	quality control
QIMA	Quality Improvement Methods Analysis
QM	quality management
RAM	random access memory
RDBMS	relational database management system
SFDC	shop-floor data collection
SMT	surface mount technology
SPC	statistical process control

SSADM	Structured System Analysis and Design Method
SQL	Standard Query Language (for databases)
TQ	total quality
TQC	total quality control
TQM	total quality management
Unix	a multi-user computer operating system
VLSI	very large scale integration (of ICs)
Z80	a type of microprocessor

Part One

Philosophy and strategy

Quality – what does this term mean to those involved in manufacturing industry? Does it refer to certain mundane monitoring activities carried on in factories by a group of rather earnest people? Or is it perhaps the key to excellence in our working lives, involving and motivating us all in a search for better ways to satisfy our customers? For many people in manufacturing, there appears to be a confusing combination of both these extremes at the centre of the quality concept. The first part of this book must therefore set the various quality themes in context.

Chapter 1 summarizes the place of quality control and quality assurance in the manufacturing system, and outlines some of the important themes in quality thinking, showing how they have come together in concepts of total quality management. Quality systems standards and their implementation are briefly discussed, and the current status of automation in the quality function is outlined.

The case for systematic automation of quality systems is made in Chapter 2, with reference to the implications of recent trends in manufacturing and their effects on a total quality philosophy.

Chapter 3 looks at the place of the automated quality system in concepts of computer-integrated manufacture. Various types of integration between the quality function and the rest of the business are proposed.

Chapter 4 takes a new look at quality performance and how it is measured. The existing methods, including defect level and cost of quality, are examined and found to be wanting. An alternative approach is proposed which should be particularly suitable for automated quality systems.

1

Quality in manufacturing

1.1 INTRODUCTION

Ensuring customer satisfaction through effective quality management is widely recognized to be a key business strategy, and fundamental to a successful manufacturing operation. In its broadest – total quality management (TQM) – sense it is widely promoted as an attitude which encompasses all aspects of the business. The most effective management approaches, technology and control systems may be applied to improve all business operations. This strategic approach to total quality (TQ) is an important theme of modern thought in manufacturing, but it must be admitted that in many cases the rhetoric is not matched by the reality.

On a more operational level, the activities required for the achievement of quality goals are of the greatest importance, and are usually the task of those in the quality function within a manufacturing business. In recent years the vital importance of the quality function has been recognized in many companies, and greater scope has been achieved for quality, with the traditional reactive monitoring and inspection roles being supplemented by a responsibility for leadership in quality improvement. Despite these advances, it is still comparatively rare for the quality function to achieve a high profile within a company, equivalent for example to the status often held by marketing.

Before going on to discuss the evolution and scope of quality ideas, it is helpful to review a few of the currently accepted definitions in the quality area. The international standard of quality vocabulary, ISO 8402 (International Organization for Standardization, 1986), gives us the following:

1. Quality management (QM) is broadly defined as 'that aspect of the overall management function that determines and implements the

quality policy'. QM is an element of long-term corporate policy. It is notable that there is no definition of TQM in the standard.

2. Quality assurance (QA) is the term used to refer to 'all those planned and systematic actions necessary to provide adequate confidence that a product or service will satisfy given requirements for quality'. As a major element of effective QM, QA covers all the specifically quality function activities of a business. The given requirements are nowadays usually interpreted as those of the customer.
3. Quality control (QC) refers to 'the operational techniques and activities that are used to fulfil requirements for quality'. As such it is a sub-set of the QA activities of the business.

Most of this book is concerned with the automation and integration of the operational and functional aspects of quality within the manufacturing plant, and hence with QC and QA. However, it is essential to review QC and QA in order to place these quality activities in the context of the manufacturing system, and of TQM ideas.

1.2 QUALITY CONTROL AND THE MANUFACTURING SYSTEM

As defined above, QC is relatively limited in scope, but it is far more than a group of inspection activities designed to ensure that defective products are not allowed to reach the customer. QC activities form a vital information feedback loop for the whole business, with potential influence on the design, process planning and logistics functions as well as on manufacture. As well as the identification of non-conformance, this feedback allows:

1. the prevention or reduction of product non-conformance, by exercising process control based on quality information, together with the rectification of any non-conformance (where possible and economic);
2. the use of process capability information in the improvement of process technology, process planning and product design;
3. the use of incoming material and component quality information to exercise control over vendor quality levels;
4. the revision of material requirements plans and work schedules to take account of non-conforming products.

QC within a manufacturing plant may be exercised by procedures such as laboratory materials testing, non-destructive testing (NDT), in-process monitoring, post-process inspection and product test. Inspection and testing may also be required for incoming goods. QC involves the recording, analysis and reporting of the quality information obtained. This information provides the basis for day-to-day decision-making on quality matters.

Many manufacturers still operate quality systems which can only be described as pure QC, with none of the wider aspects described below.

1.3 QUALITY ASSURANCE

As well as QC functions, the broader scope of QA also embraces the assurance of designs, manufacturing processes and customer service. The planning of QC procedures and the evaluation of their effectiveness using auditing procedures are QA activities.

QA systems are management systems which develop and implement aspects of the defined corporate quality policy. Procedures are laid down to assure that quality decisions are made by the appropriate person, using the official guidelines, and on the basis of the best possible information. QA systems are often designed and formally documented in accordance with national and international quality systems standards.

Several important themes can be identified in the development of modern QA thinking. The earliest is the use of statistical and probability theory.

Statistical process control (SPC) was initiated in the 1930s by W. A. Shewhart, who suggested the use of control charts as a tool to monitor and control production processes. After many years in the wilderness, SPC techniques have now been recognized as one of the major tools for the control and improvement of quality, not only in manufacturing but also in administrative and service functions.

Some years after the development of SPC methods, the application of probability theory in the development of sampling techniques for the economic quality assessment of large batches was first described by Dodge and Romig (1941). This approach has found major application in the quality assessment of incoming goods, and the control of supplier quality. A great deal of research on the development of these basic ideas has been done, reflecting the interest of mathematicians in such practical applications of probability theory.

However, since sampling requires the definition of acceptable quality levels, an anathema to zero defects advocates, in recent thinking on supplier relationships and defect levels these methods have lost favour.

Later, the work of W.E. Deming and J.M. Juran, the first of a species sometimes known as 'quality gurus', beginning in the 1940s, was much concerned with the development of a systematic overall management strategy for the assurance of quality. Many concepts in the quality field were first expressed by one of these men, who with their long and influential careers have ensured that their ideas are firmly established in industry and form much of the basic thinking in quality.

In recent years the important human factors themes developed by management thinkers have permeated the quality area. It has been realized that many quality problems can be tackled by a different approach to the management of human resources, aligned to a 'theory Y' management style. Morale, motivation and a commitment to quality may be fostered by a management approach which encourages employee participation in quality problem-solving. At the same time it has become obvious that the best inspection system cannot compensate for a careless and poorly motivated production worker. Quality cannot be 'inspected in'.

A typical example of emphasis on human factors themes is found in the introduction of quality circles, specially organized teams of workers who endeavour to identify and rectify quality problems. These have been notably successful in Japan, and attempted with varying results in other countries. The adoption of such approaches has had the beneficial effect of helping challenge outdated management styles in many companies, but, to quote Sinha and Willborn (1985), 'a real breakthrough for quality management has not been achieved as a result of these programs'.

1.4 TOTAL QUALITY MANAGEMENT

The ideas of TQM and total quality control (TQC) have arisen from a synthesis of elements from the QA themes above, together with a widespread appreciation that a more all-embracing approach is needed to ensure competitive quality performance. There are a number of leading pioneers and proponents of TQM and TQC ideas, notably Crosby (1979) and Feigenbaum (1983). Feigenbaum identifies three main elements of TQM: tools and techniques, quality systems and human factors. These correspond to the themes of quality thought reviewed above, and emphasize the fusion of previous themes within TQM.

There is a plethora of different definitions and models of TQM, many developed by academics. They usually agree on certain central elements, but there is still a substantial amount of confusion and discord about emphasis. TQM is seen as a process by some, as a philosophy or structure by others. According to Newell (1990), who has studied a number of companies in different stages of the introduction of TQM, many managers are uncertain about the distinction between TQM and a quality improvement programme, and most companies are unable to identify, plan or measure stages of progress towards TQ.

What is widely realized, however, is that a TQM approach requires a change in the organizational culture, in which a number of core principles may be identified. They include:

1. The replacement of mass inspection with a systematic, process-orientated approach to the prevention of quality loss at the earliest possible (preferably the design) stage, through a process of never-ending quality improvement and the elimination of waste.
2. Motivation, mainly through involvement (often in groups or multidisciplinary teams) of all employees in the company-wide process of never-ending improvement using the concept of the 'internal customer'.
3. Training in the use of statistical tools, in the identification and solving of problems, and in improvement techniques to allow quality information to be obtained, evaluated and used by all. Effective measurement and analysis are considered to be fundamental to a TQM approach.
4. A simplified working environment, without slogans and over-bureaucratic imposition of work standards, to encourage innovation, openness and good communications. The setting of realistic quality objectives and bench-marks for performance is, however, still vital.
5. Most fundamentally, the creation of a responsive corporate culture which is orientated towards the market, centred around the needs of the customer, and driven by sincere commitment from the highest levels.

The implications of a TQ approach are widespread, and affect many aspects of an organization. TQM ideas are associated with other recent developments in management thinking concerned with the improvement of business effectiveness, which together constitute a move away from Taylorism and 'theory X' management.

Comparisons have been made between workforce attitudes and industrial cultures in the various industrialized economies, and Japanese success in quality matters is held to be an example of what can be achieved by the involvement and motivation of a receptive workforce in a TQ approach.

It is interesting to note that some manufacturers have felt it necessary to make a clear distinction between the product quality assurance system, which they see as a contractual obligation to their customers and as the responsibility of the quality function, and their effort to embrace TQ concepts, which is viewed as the search for excellence and is the responsibility of everyone. In some cases a manager is appointed with particular responsibility for an improvement programme.

1.5 ZERO DEFECTS

One objective of many manufacturing companies which have adopted TQM ideas is 'zero defects' – often interpreted to mean the total elimination of non-conformance. There are a number of different varieties of the zero defects approach; for example the 'poka-yoke' system described by Shingo (1986), where the improvement process is concentrated on 'mistake-proofing' the manufacturing process and equipment. Shingo provides a number of examples, including that of an assembly section employing 23 workers, and producing 30 000 complex sub-assemblies per month, where no defective sub-assemblies were manufactured in a period of six months.

Zero defects approaches are usually based on the traditional concept of product specifications having a nominal value with associated tolerances. Some theorists make a distinction between zero defects systems based on tolerances, and a policy of using never-ending reduction in process variability - aimed at the nominal value – as a criterion for excellence. The latter is sometimes regarded as a superior form of quality improvement, leading to greater customer satisfaction.

The vital importance of a policy of never-ending improvement aimed at process and product cannot be doubted, but the ability of such an approach to eliminate non-conformance totally has not been satisfactorily demonstrated for every manufacturing environment. In fact, graphed data provided in studies such as those described by Schneiderman (1986) and Mills (1987) seem to demonstrate (contrary to the intention of these writers) that the law of diminishing returns applies as defect levels are reduced to a level undoubtedly much lower than it was initially, but which is not negligible and shows no tendency to reduce to zero over a period of years.

Zero defects is seen as a realistic objective by some, and has found favour among quality improvement practitioners, consultants and academics. However, many practical manufacturing and quality engineers would agree that process improvement and even 'qualification' of an established fully capable process can never offer complete protection against an unexpected event which degrades product quality. In addition, many manufacturing processes have an inherent variability which defies qualification. In these cases active in-process or post-process QC procedures must remain a crucial element of quality management strategy.

1.6 IDEAS OF QUALITY AUTOMATION

There are two aspects to the automation of quality activities using information technology (IT). First, there are the concepts – the strategy for automation. Most manufacturing engineers and managers will be aware of existing automation strategies for the automation of the design (CAD), manufacturing (CAM) and production management (CAPM) functions within a manufacturing system. These strategies are sometimes combined in a grand design known as computer-integrated manufacture (CIM).

Second, there is the practice – hardware and software systems based around the strategies for automation. Without a coherent and accepted strategy for the automation of any manufacturing function, it is difficult to develop – let alone sell – an overall solution. At the time of writing, this is the position in the quality function. The current state of quality automation practice will be briefly outlined below, and the technology examined in Part Three. First, we should outline the concepts and strategies for quality automation which have been proposed in recent years.

The terms 'integrated quality control systems' (IQCS), 'integrated quality systems' (IQS) and 'computer-aided quality' (CAQ) have developed a limited currency, and basic requirements for automation in quality systems have been described by a number of people, both academics and quality practitioners. A brief survey of typical ideas is given below.

In *Computer-Integrated Manufacturing Technology and Systems*, a standard book on CIM, Rembold *et al.* (1985) are almost unique among CIM writers in giving quality aspects a lengthy treatment, concluding that 'In future manufacturing facilities, hierarchical computer systems will be employed to supervise an integrated quality control system', but that 'To date, not all the building blocks for an integrated quality control system are available'.

An approach to the use of computer technology to provide 'quality tools' at AT&T, a major multinational manufacturer, is described by Ackerman *et al.* (1986). In the AT&T approach, the four major themes are in-process control, process quality, the ability to analyse and make use of variable instead of attribute data, and 'company-wide quality management'. It is the evolution in thinking in these areas which is having an impact on the use of computer technology in quality at AT&T. There is an emphasis on real-time process control and on the use of quality information for improvement, but no fully developed concept of quality systems integration with other manufacturing functions.

Willborn (1986) examines the effects of automation of the manufacturing process on QC systems designed for human-dominated production systems, and describes the applications, advantages and likely impact of the use of computers for inspection and quality monitoring. He also identifies possible integrations with other business and manufacturing functions, and concludes that automated QA will become an integrating bond and coordinating force in the organization. The idea of a quality database and some possible contents are also described.

A definition of the IQS and a method of systematic quality system design have been proposed by Dessouky *et al.* (1987), using the concept of manufacturing process transfer functions with related 'quality control windows' performing the functions of a control cycle. These building blocks are to be arranged within analytical frameworks within which the information flows can be described. This interesting approach is meant to operate at all stages of a product life cycle, and might be described as the control engineering approach to quality, in that it utilizes concepts from that field. There are suggestions that automation of the in-process QC function should be associated with integration.

Another broad concept of CAQ is described by Tuttle (1987), who examines requirements for an integrated system incorporating a comprehensive set of quality functions. He suggests that the implementation of such an approach would provide a strong competitive advantage.

From this brief survey it may be seen that the ideas which have been proposed for quality automation tend to be rather strategic and theoretical, lacking the level of detail which would be necessary for implementation. The detailed systems descriptions which are available usually refer to individual items of quality data collection equipment, but these are not usually conceived in the context of an integrated system. There is something of a credibility gap, then, in quality systems automation which it is the purpose of this book to fill.

1.7 THE INFLUENCE OF QUALITY SYSTEM STANDARDS

Many quality managers are closely involved with quality systems standards. In recent years the methods by which quality in design, manufacture and services should be managed have been formalized and standardized, starting with the requirements of defence procurement agencies in the United States. Most industrialized countries now have a civil standard (which has evolved from the military standards) for quality systems, prepared by the national standardization body. In Britain this standard is BS 5750 (see Appendix C). Like BS 5750, most of these standards are now aligned with the international series of standards for quality systems, the ISO 9000 series (International Organization for Standardization, 1987).

A company having a quality system which conforms to the requirements of the appropriate standard may gain considerable competitive advantage because of the superior control over quality matters and documentation which a comprehensive and logical and management system allows. The systems produced are concerned with QC and QA, and do not require, for example, that a company adopts a TQ approach. A process of never-ending improvement should be facilitated by the adoption of a standard, but it is not a stipulation. Quality costing is another aspect not covered by such standards, although some of the original military standards did include this area.

In order to achieve a system which meets the requirements of the appropriate standard, QA policies and procedures must be devised and documented. It is extraordinary to find, as is frequently the case, that the system which is designed is largely or completely paper-based. Forms are devised, filing cabinets labelled and filled with records, and card indexes produced.

If a new systems standard for wage payment were to be introduced, we would not expect the finance function to develop a paper-based system, and our wage and salary slips to appear again as manually completed documents on a pre-printed form. We would assume that data processing or computer systems analysts would be brought in to design and implement an automated system able to meet the requirements of the standard.

Why do so many quality managers come up with such traditional arrangements? The answer, perhaps, lies in a poor appreciation of the potential advantages of IT automation in the quality area. The quality function is not renowned for computing awareness and knowledge. The greatest problem, however, is the lack of a coherent strategy and framework for automation. There is no reason why the requirements of quality systems standards cannot be fulfilled by computerized systems, using suitably designed automatic methods for quality data collection, analysis and record keeping.

Computer systems analysis methods have their place in quality systems design, but the importance of human factors in the quality area really demands a technique which is capable of effectively integrating people and machines. In Part Two, a structured design and improvement method for quality systems which facilitates the introduction of automated and integrated systems will be introduced, to show one method for achieving this integrated design.

1.8 QUALITY AUTOMATION IN PRACTICE

To conclude this introductory chapter, it is useful to outline the current state of quality automation practice in manufacturing. Part Three will examine the technology in greater depth.

Perhaps because of the lack of coherent guidelines for quality automation and integration, the majority of manufacturing organizations have as yet made little progress in this area, although they will be using computers in data processing for other purposes, and probably also for design, material control and production scheduling. An early survey in Britain by Opdebeeck and Oakland (1983) indicated that a very low proportion of manufacturing companies were then making use of computers in their quality systems. No more recent studies are available, but a large number of software products have come on to the market since that survey; these are mainly designed to perform SPC functions, by the automation of control charting and process capability studies from keyboard input of data. It may safely be assumed that many more companies now make use of QC software, using mainly stand-alone SPC packages running on microcomputers.

If the author's observations during regular visits to manufacturing plants of all types is to be trusted, there has also been a great increase in the use of more sophisticated approaches, such as the computerized shop-floor dimensional gauging systems available from metrology equipment suppliers. These usually accept direct input from sensors applied to components. The computer then indicates conformance to specification, and calculates SPC information for process control. These systems tend, in practice, to be inflexible in application and incapable of integration with other systems.

Automated in-process QC techniques are also advancing, and are found in a wide range of manufacturing industry. There is a distinction between process control, where analysis of a process variable is used to control the process, and in-process product inspection, which can often provide more valuable quality data.

The important technology of coordinate measuring machines (CMMs) provides perhaps the most advanced and flexible equipment available for post-process dimensional inspection. Machine vision and automatic test equipment (ATE) also represent major themes in this area.

A number of computer and software suppliers now offer complete quality information management systems, based either on multitasking minicomputers or networks of microcomputers. Cullen (1987) has identified and described four typical systems. Among the most comprehensive in scope was Hewlett-Packard's QDM package. This product was initially developed for a Hewlett-Packard disk drive manufacturing plant and later made available as a software product.

QDM includes a database system with report generator, and allows input of quality data from a wide range of automatic and manual sources. Special attention was paid to ease of configuration of the system to accept data input from non-standard sources. Users had considerable freedom to develop their own data acquisition and reporting applications centred around the QDM core. One use of this system, combined with other software and with ATE, at a Plessey manufacturing plant, is described by Dowding (1987). Excellent results were reported, and the total system has many advanced features, including a certain amount of integration with production planning computer systems.

These, then, are the main themes in quality automation practice: SPC software, automated inspection systems, in-process quality control, and integrated quality information systems. Suppliers of the first two types of system are tending to develop their products towards fully integrated systems, usually by the addition of database functions. A further development is a trend for shop-floor data collection (SFDC) systems (currently used mainly for 'quantity' data) to include a limited quality data capability.

1.9 SUMMARY

This chapter has attempted to cover a great deal of ground in an effort to describe the background to quality systems automation. The scope and place of QC and QA in the manufacturing system have been outlined, and a look taken at the involved area of TQM, where some core elements have been listed. The thorny question of 'zero defects' has been discussed briefly, and the conclusion reached that despite zero defects approaches to process improvement, QC (either in-process or by post-process inspection) is here to stay.

The current state of the art in quality systems automation was examined, with a review of some of the emerging concepts and existing technology. The influence of quality systems standards was discussed, and the lack of a trend towards systems automation deplored.

The conclusion which must be reached is plain. Automation in quality systems lags a long way behind other areas of the manufacturing system, particularly in terms of accepted concepts and strategy for automation. The reasons for, and effects of, this situation are discussed in the next chapter.

2
Why automate?

2.1 INTRODUCTION

Many of the main themes in current philosophies of QA were developed before the advent of affordable computing power and now appear out of step with recent developments in other areas of manufacturing, where the elimination of stocks, flexible automation and rapid response are at a premium. Effective automation in quality data management should be seen as an essential integrating mechanism in any move towards CIM. Quality function activities cannot continue in their present form into a future where manufacturing systems may be otherwise largely automated. Quality must not become an 'island of paperwork' if flexible and efficient manufacturing is the goal.

The traditional approaches to TQM are widely accepted as worthwhile, and indeed essential, but effective quality management also depends on the timely provision of adequate and correct information to support a process of never-ending improvement in the business at all levels, from process operations to senior manufacturing management. Data integrity is vital for efficient operation in the modern automated manufacturing environment, and this means in practice that the introduction of effective integrated computer systems in quality is essential, as is already the case in many other areas of manufacturing.

2.2 JUST-IN-TIME AND QUALITY SYSTEMS

Just-in-time (JIT) is one of the most important themes in modern manufacturing. There are many different definitions of JIT, some of which are philosophies very similar to the TQM concepts which emphasize the elimination of waste, while others concentrate on tools and techniques for the elimination of stocks. Most people agree, however, that the adoption of JIT will place the organization in a vulnerable position if quality problems occur, as there are no safety stocks to fall back upon.

An emerging problem, as many manufacturers turn to JIT, is the reaction-time limitation of the traditional paper-based quality system. When attempting to tackle a modern high-variety and low-stock manufacturing environment, the time taken to complete the quality information feedback loops means the system may be unable to cope with demanding products.

Long intervals between a failure event and the resulting report make it harder to reach agreement on corrective action, while the summary reports which are normally all that can be provided smooth out the short-term variation which might have provided vital information for defect prevention.

Graham (1988) states that the communications links connecting inspection with other manufacturing functions form a serious source of inefficiency in JIT manufacturing environments. He emphasizes that the importance of rapid quality information feedback systems is being increasingly recognized as a means of achieving significant improvements in quality levels.

Automated real-time quality data collection and evaluation is the solution, allowing effective process improvement to occur.

2.3 DATA VOLUME

There is a trend towards automation in quality data collection, made possible by new technologies available, such as CMMs, ATE and machine vision. This tendency is sure to result in a large increase in the amount of product and process quality data available. Dowding (1987) reports that up to 500 000 measurements per day were being performed at a Plessey electronics manufacturing plant after the introduction of ATE. In many plants it is possible to find stand-alone CMMs and other expensive inspection systems, the detailed output from which is clearly under-utilized, as there is no time to collate and analyse the data.

Product quality information is of great value. If properly used it allows effective control and improvement of the manufacturing function. Some efficient method of assimilating large volumes of data is vital, in order to create an environment where the correct quality information is made available to the appropriate person at the right time, and in which automatic feedback of control information to the process may be used to improve quality levels.

It is not sufficient to automate only the collection of quality data. The management of quality information must also be tackled. This involves not only storage, but also analysis, correlation on a plant-by-plant or company-wide basis, and reporting.

2.4 HUMAN FACTORS

Any suggestion of the radical extension of IT automation to the quality function will meet with disagreement from certain quality practitioners, whose main interests lie in the human aspects of never-ending improvement programmes. This they rightly see as fundamental to successful TQM. Their greatest concern is that the introduction of quality automation will prejudice the valuable team spirit – the sense of responsibility for and control over quality – which they believe they have developed in their workforces. By implication, they associate automation with 'theory X' management and a deskilled, demotivated labour force.

While hard, inflexible manufacturing automation has deserved such connotations, the proposed automation is not of that variety. IT automation has not, as expected by some, led to total control from the top. On the contrary, flexible automation and IT frequently do not flourish in a traditional organizational setting, and may in fact tend to subvert it, chiefly by threatening the role of middle management. This role, based on the gathering, processing and collation of information, and its communication between strategy-makers and the productive workforce, may be rendered largely redundant by IT.

The real experience of IT automation in many more enlightened organizations is that it leads to the development of strong teams of technicians, highly skilled and motivated, and capable – unlike their managers – of reprogramming the computer systems in their charge. These teams are most effectively managed with weak control structures, rather than the strong control systems associated with traditional hierarchical management structures. Attempts to manage them in the traditional manner can be ineffective, and can even expose a lack of up-to-date technical ability in their superiors.

Hence, automation in quality systems should be a support to the human factors themes of TQM, and, rather than undermining teamwork, it will emphasize the importance of strong self-managing teams, dedicated to high quality in their working lives. In addition, it will provide them with a far superior level of quality information, and free them from many time-consuming routine tasks of quality data collection and analysis.

2.5 THE BENEFITS OF QUALITY SYSTEMS AUTOMATION

The benefits of the introduction of this type of quality system for the organization should include:

1. Improved support for quality decision makers at all levels of the organization by the rapid and correct provision of quality information, enhancing the effectiveness of a process of never-ending quality improvement.
2. Increased product value and customer satisfaction, resulting from improvement in outgoing product quality levels, caused by the use of improved product and process QC methods.
3. Reduction in the costs associated with manual QC systems by reducing appraisal and failure costs, mistakes, and the disruption caused by non-conformance. Another aspect is the elimination of paperwork and manual record-keeping, with its associated costs and errors.
4. The ability to cope with flexible automation and low-stock manufacturing organization by reduction of the feedback time delay associated with paper-based systems.
5. An enhanced ability to make operators and supervisors directly responsible for the quality of their products by providing them with immediately available and correctly structured information on process performance.
6. A strategy for future integration into CIM systems, based on a systematic design methodology, together with consideration of desirable interactions with other manufacturing sub-systems.

In a revealing case study of a Pirelli tyre manufacturing plant, Oliver (1986) describes the introduction of an automated quality data collection and analysis system. The system includes a minicomputer holding process set-up and quality parameters, linked to local microcomputers monitoring the rubber mixing process and providing SPC charts for the analysis of trends. The resulting benefits were clear. Process set-up (especially for new products) was eased, as process behaviour could be more easily examined. This was initially considered to be the main purpose of the system. Oliver also reports that:

> Another (unintended) consequence of the system is to provide more leverage to those responsible for overseeing the process. Whereas corrective action before computerization had been driven heavily by hunches about what the process was doing, the IBM can produce hard data about the performance of aspects of the process. The ability of the process controllers to produce a printout 'proving' their point has led to swifter responses from maintenance teams and other groups and individuals who are called in to rectify problems.

In addition to the storage of process set-up and quality parameters, quality defect information is entered directly on to terminals linked to the central minicomputer, eliminating manual recording. This means that collated and analysed information is available much more quickly. Previously quality data were analysed for consideration on the following day. With the automated system in operation, shift supervision and management can be provided with reports and analyses at any time.

Pirelli views its quality automation as a success. The system has resulted in estimated savings of 0.3% on scrap and 0.5% on other non-conformances (such as appearance faults). With a production rate of 6 000 high-value tyres per day, these savings are substantial.

Another example of the benefits of automation is provided by Withers (1990) in an article describing the approach taken by Schlumberger Industries Flow Measurement, a company manufacturing gas meters – products for which quality and safety are paramount considerations.

Unlike Pirelli, Schlumberger purchased a purpose-designed quality management software product, QMS, from Strategic Systems International, which was introduced in an effort to provide better information for quality improvement. The software runs on a minicomputer, and can be accessed via 60 shop-floor terminals which are used for data entry and reporting results.

QMS is a database system which must be customized to suit the user's needs, having no standard data entry forms. This configuration process took 18 person-months, with substantial support from the system suppliers. The system functions cover the collection, storage, analysis and presentation of quality information from all inspection and test stages. It eliminates the need for inspection paperwork, and provides process capability studies, SPC, calibration and defect management (the last aimed at supplier performance monitoring).

The results were impressive. Quality information was much more readily and rapidly available for improvement and the prevention of quality loss. The rate of reject units fell by 2% overall, about one-third of the original reject rate, in one year. In the same period production volumes rose by 40%. These savings were sufficient to pay for software and development costs several times over.

Future development plans for the system involve the replacement of manual data entry by the use of automatic test data collection equipment, and interfaced hand-held inspection instruments. This is part of a programme to improve the flow of data to the QMS system, which also includes the introduction of barcodes and light pens.

More important than cost savings, perhaps, is the far greater awareness of quality issues that the introduction of QMS has brought to many at Schlumberger, and the opportunities it allows for QM to concentrate on the planning of improvements and future developments, now that the basic data collection and analysis task has been automated.

2.6 SUMMARY

This chapter has examined the rationale for the automation of quality function activities, and the benefits which should result.

The reasons for automation are found in three main areas: first, the need to accommodate new patterns of manufacturing activity, such as flexible automation, CIM and JIT, which have meant that accurate information must be immediately available to allow the identification of quality problems; second, the great increase in the volume of quality data provided by the new technologies for quality data collection, which demands an effective quality data management strategy; and finally, and most importantly, the need for timely and accurate information to aid the quality improvement process.

It is here that automation has a key role to play in supporting a TQ approach. By relieving quality and manufacturing personnel from the routine tasks of data collection, analysis and reporting, it allows them to concentrate their efforts on the improvement process, which is the path to excellence in manufacturing.

Six important benefits are identified in all, supported by two examples which demonstrate that cost savings and improvements in efficiency may be gained from quality systems automation.

3

The integrated quality system in computer-integrated manufacture

3.1 INTRODUCTION

Computer-integrated manufacture (CIM) is an area where it is still useful to indicate the scope of the concept of CIM in use, before looking at the detail.

Early ideas of CIM tended to be little more than an integration of the functions of the computer-aided design (CAD) and computer-aided manufacture (CAM) sub-systems. In omitting the other interactions of the manufacturing function, this approach failed to recognize that the benefits of CIM only arise from the ability of an effectively integrated manufacturing environment to respond flexibly, economically and rapidly to the market.

CIM is not limited to the transfer of data between CAD and CAM sub-systems, but incorporates these aspects within a wider systems view to indicate the integration of the manufacturing, marketing, finance and other business functions. CIM design is not complete with consideration of design and production, but also includes QA and other activities such as delivery, and may extend into the supplier chain. CIM should be seen as a philosophy embracing all aspects of the business, the concept then becoming that of the 'computer-integrated business'. At its widest, CIM may be seen as 'a management philosophy that allows us to optimize our productive resources' (Vail 1988).

Such a definition encompasses all possible aspects, but is unfortunately too general to give any framework for the practical consideration of CIM implementation. This definition might be applied to JIT or even TQM! The definition preferred for the purposes of this book was provided by Rzevski (1987), who referred to 'a system whose aim is to add value to the manufacturing business by employing Information Technology with a view to achieving an effective integration of all planning and control activities within the host manufacturing organisation'. The emphasis on planning and control of the manufacturing organisation is particularly useful when we are considering quality systems.

Whatever the definition adopted, a fundamental objective of any CIM system must be excellent quality performance. Without the ability automatically to monitor and control the quality of manufacture a CIM system will be incomplete, as manual data management will be required on a substantial scale in one of the major business control activities. In Chapter 2 the problems resulting from the operation of conventional paper-based QC systems in an advanced manufacturing environment were described. An automated and integrated quality system must be considered as an essential approach to tackling the problem of quality in CIM.

But how is quality to be integrated? The present chapter addresses this question – the role of the IQS in CIM – by reviewing some of the concepts and implications of integration. Necessary and desirable aspects and levels of integration for the quality function are then proposed.

3.2 QUALITY IN CIM

Quality as a topic within the study of CIM has on occasion been omitted in a remarkable manner. For example, the first CIM project funded through the European Community research programme Esprit, as described by Yeomans (1987), defined no fewer than 24 separate sub-systems of CIM without including quality. Other writer such as Ranky (1987), emphasize the importance of quality systems in CIM, as a high-priority aim of the organization, an integral part of the real-time control system and an integrated part of the design for manufacturing activities.

QA is a broadly defined concept encompassing all those functional activities of a business which affect the customer's satisfaction with the product or service in the cost/value nexus. It is defined at a level of generality similar to our preferred definition of CIM. As suggested earlier, QC is the more restricted term used to represent those operations and procedures which maintain and improve product quality levels, and is a sub-set of QA activities.

It is the operational activities of QC which must be largely automated within CIM if the full benefits of flexible and economic automation are to be realized through competitive advantage. The objective must be to develop quality systems which can take their place in an evolving CIM environment alongside the existing elements such as CAD, CAM, and computer-aided production management (CAPM).

'Integration' is a concept which requires careful consideration. It may be viewed in various useful ways. Some authors, such as Below (1987), focus on hierarchical levels of integration resulting from correct commonality. Below identifies seven levels of integration: physical (or interfacing); data; work schedule; business function (or practice); attitude; principle; and purpose. Others refer to 'direction' or 'dimension' of integration – for example, Dessouky *et al.* (1987), who suggest that two dimensions of integration should be identified arising from integrated QC systems: process-wise and resource-wise integration. These two views of integration are complimentary rather than contradictory and both are incorporated in the following suggestion for quality systems integration.

Three distinct dimensions of quality systems integration are proposed:

1. vertical integration – within the quality system;
2. functional integration – with other CIM elements;
3. process integration – through the production cycle.

Figure 3.1 illustrates the quality system integration space. In practice, the level of integration which is appropriate and possible will depend on the dimension of integration. Each of these dimensions of integration will have an associated desirable level of integration for any quality system.

Integration on the physical level (such as computer communications) and a considerable degree of data integration between system elements are certainly necessary for all dimensions of integration. These levels may be sufficient for the vertical integration dimension, but a degree of integration on the level of work schedule and business function is also needed for functional and process integration. The implications of each dimension of quality systems integration are outlined below.

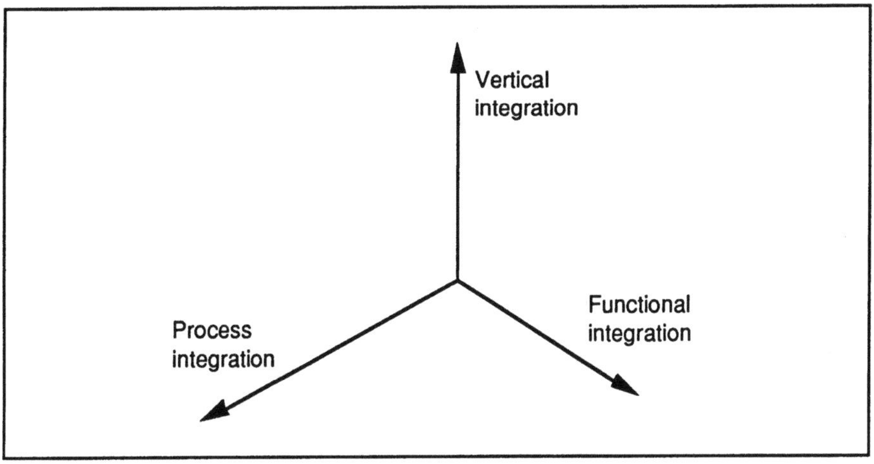

Figure 3.1 The quality system integration space

3.3 VERTICAL INTEGRATION

Vertical integration occurs within each CIM sub-system, and describes the information processing and communications abilities needed to enable the information created and held by the sub-system to be used effectively. Information may be passed upwards and downwards in this vertical dimension. The upward flow of information may be from activities such as inspection and process monitoring at the lowest level, perhaps to a quality database, and hence through to management information for decision support. In the downwards direction the information requirements of inspection activities may be made available from the quality database. These will consist of inspection plans and specifications, against which the process and product monitoring and inspections will be carried out. Practical aspects of vertical integration will be discussed in later chapters.

3.4 FUNCTIONAL INTEGRATION

Automation of the quality system has implications for integration with other functional elements within a CIM. The possible and desirable connections with these elements can be identified by considering the uses of quality information by the other sub-systems, and the information needs of the quality function. Physical integration using computer communications is assumed, although it is not always an insignificant task to implement in practice. Most of the connections suggested below are on the level of data integration.

3.4.1 *Product design*

The activity of product design produces product characteristics with associated nominal and tolerance values which may be automatically interpreted for use in the quality system as well as in manufacture. This requires the processing of design data (perhaps from a CAD system) into a form which can be stored in a database of product quality specifications and downloaded for use by automatic inspection and test equipment.

Data can also flow in the other direction. For example, it is well known that tolerances which are unnecessarily tight may cause production difficulties resulting in increased costs. Automatic feedback of data to the design department, concerning those product characteristics which are consistently difficult to manufacture within tolerance, will encourage designers to be more aware of process capability limitations and help reduce these costs. The relevant process capability information can be made available from the quality system, and should ideally be accessible from within the CAD system.

3.4.2 *Manufacturing process design and planning*

The design and planning of the manufacturing process has important implications for product quality. Both initial process selection and detailed operation planning should be made on the basis of the best possible information about process capability as well as cost. There is a move towards the use of computers for the automation of manufacturing planning. Quality information will be a vital input to such systems, as well as to the existing manual systems.

Product and process design, together with process planning, will require a higher level of mutual integration in the future to improve design-to-manufacture performance. This integration will have to exist on the level of business function rather than the work schedule integration which now exists in well-organized manufacturing companies. Such integrated systems are referred to as 'concurrent' or 'simultaneous engineering', and are an important aspect of integration for CIM. Simultaneous engineering will require quality-related information concerning supplier quality as well as process capability.

3.4.3 *Manufacture*

The automation and integration of QC clearly has vital implications for the manufacturing function. Obviously the identification and diagnosis of non-conformance is critical, but the most immediate feedback loop from the quality system is that which allows the adjustment of the manufacturing process to *prevent* non-conformance, by the use of process and product quality information.

The same quality information can be analysed to enable manufacturing process improvement. But if the improvement process has yet to eliminate all significant variability in the process, active control – performed either manually or automatically – will still be required.

There are three main ways in which quality information may be obtained: by the monitoring of process variables, and by in-process or post-process product inspection. Automated or semi-automated process control systems, where the control information is provided by one of these methods, are in use throughout manufacturing industry, and, of course, there are few manual processes which cannot benefit from immediate feedback of quality data to the operator.

In most cases the process quality data collected are not used for any purpose beyond control of the process in which the measurement activity takes place, but such information can also be used to provide process control information for a preceding process, and indeed for a subsequent process. The latter is known as 'forward control'. The implications are discussed in Section 3.5.2.

It is only the use of local quality monitoring and process control systems which can reduce the scale of the central real-time processing and communications requirement in a CIM system to viable proportions. Integration between the quality data collection elements and the manufacturing process controllers should normally be at the lowest possible level, to avoid unnecessary vertical transfer of quality and process control information.

3.4.4 *Production management*

Immediate automatic feedback of the number of non-conforming items at each stage in the manufacturing cycle makes possible the immediate updating of work schedules and materials requirements. This should reduce the need for costly and disruptive small batches to be scheduled compensating for non-conformance notified after a works order has been completed. More efficient material requirements plans may be produced, reducing inventory management costs.

At present only a few CAPM systems are capable of real-time schedule update, but many software houses are beginning to move in this direction. Developments in this area are linked to the increasing use of SFDC as a source of real-time data concerning quantity, and – to a limited extent – quality.

CAPM systems such as MRP, MRPII and OPT represent a large investment for a manufacturing company, and despite the current popularity of low-tech JIT approaches (such as *kanban*) to production control, it is difficult for many companies to see a way of eliminating their dependence on such systems. The most advanced companies have realized that there is no need to choose either one approach or the other. Different types of product require different production management solutions, which may exist side by side in the same plant. As hybrid CAPM systems are developed to enhance flexibility and responsiveness, improved integration with the quality function, via real-time feedback, will be a necessity.

3.4.5 *The business system*

To realize the full potential of computer integration in the business, QC – as an important feedback system – must be integrated with such business functions as logistics, customer service, finance and strategic planning.

Integration with logistics is vital, to allow effective control over the quality of purchased items. The acquisition, collation and analysis of supplier quality data is a central activity of a quality system, and the purchasing function will require regular access to this information, which should be available in a format compatible with its supplier database.

Customer service, for many manufacturers, forms an important source of product quality data which should not be neglected. Complaints and warranty claims, together with reports from service engineers, dealers and franchise holders, constitute sources of information close to the customer. If this information is made available in the correct form it can be collated with quality information resulting from product inspection.

The finance function in manufacturing industry frequently takes little account of quality costs. However, information concerning the cost of products at various stages of manufacture is often available. If non-conformance data are readily available in an appropriate format, the importance of investment in quality improvement may become more evident to the 'bean counters'.

Quality performance information available to senior management charged with strategic decision-making should be of a general nature, but should illustrate the financial implications of quality loss and the importance of appropriate investment. The use of relevant and sensitive criteria for the assessment of quality performance is essential, and such measures are examined in detail in Chapter 4.

3.5 PROCESS INTEGRATION

The final axis of integration which requires consideration is that of QC through the cycle or sequence of manufacturing operations for any product. Within the quality sub-system, process integration involves two aspects: first, that concerned with the planning of inspection and monitoring activities; and second, with the use of quality data for process control at a different stage of the manufacturing cycle from the stage at which it was obtained. This might be termed 'remote process control'.

3.5.1 *Inspection planning*

Inspection planning consists of the creation of inspection specifications and plans for use in manual, computer-aided and automatic quality data acquisition, using as a basis the product and manufacturing specifications.

Inspection and process monitoring are the source of essential information which is useful throughout the business, yet little formal attention has been paid to the effective planning of these activities to obtain the best possible quality of information. Such planning as is performed usually forms part of the manufacturing planning process. It is commonplace to find detailed manufacturing specifications followed by the word 'inspect', presumably referring to the last operation, with no indication of procedures or instruments.

With the use of automated inspection and process monitoring techniques, this cavalier approach to inspection planning must be replaced by more formal methods. This involves the identification of each element of the manufacturing cycle, together with the features which require inspection or monitoring at that stage, the selection of an appropriate monitoring method and station, and the passing of the correct procedures and specifications to that station. An automated system to perform this planning activity might be termed 'computer-aided quality planning' (CAQP).

The development of inspection specifications and programs from CAD systems is presently achieved (where it is attempted) through interactive sessions in a manner similar to computer-aided part programming, where component features are identified by the programmer, and suitable equipment or techniques nominated to perform the task. Where computerized CMMs are employed, standard post-processing of the interactive session output can provide programs to control the machine through its required inspection task. The Dimensional Measuring Interface Specification (DMIS), developed in the USA, is designed to be an industry-standard language to communicate process-orientated commands and geometric definitions from a CAD system to any computerized inspection device. At least one software product is available to generate CMM programs using an interactive method from CAD data.

The use of artificial intelligence (AI) techniques for inspection program generation offers one avenue for exploration; a considerable amount of research work is taking place in this area. AI systems are also called 'knowledge-based systems' or 'expert systems', and represent a new and fascinating direction in computing, which attempts to emulate the methods of the human brain.

In one project currently under active development at a major aerospace manufacturing company, detailed inspection programs for each component will be generated using an expert system. This system is called an 'intelligent inspection planning assistant'. The programs will be downloaded to computer-aided inspection (CAI) workstations situated on the shop floor, where they will be used to guide the user through complex quality assessments. This will mean that the company will be able to ask personnel such as semi-skilled process operators to inspect and certify their own work. The CAI workstations collect the quality data directly from interfaced measurement instruments for collation and analysis. (Chapter 14 discusses the CAI concept in more detail.)

Initially, inspection procedures for each type of feature are developed and validated by expert quality engineers. This information forms the knowledge base, consisting of special 'object' representations for each quality feature with associated 'rules'. Rules are also obtained by observing and interviewing expert inspectors to find out the best methods for inspecting each type of feature. Rules are simply ways of representing knowledge and expertise. For example, a rule might state 'If feature is bore and diameter is greater than 250 mm and length of bore is greater than 100 mm, then measure diameter using internal micrometer at three positions along length of bore'.

The inspection planning system user will pick the type of feature to be inspected from a menu, and add supplementary information concerning nominal dimensions and tolerances. The system uses its rule base to select instruments, ancillary equipment and measurement procedures and provide user instructions for the CAI workstation. On-screen graphical displays will show the feature to be inspected. Each stage of the manufacturing process will have the appropriate inspection activities provided by the program, concluding with the final view, where every feature is re-examined.

3.5.2 *Remote process control*

This is the use of quality information derived from one stage in the manufacturing sequence to control another stage, in order to avoid non-conformance in the product. This can be achieved by two methods, known as 'backward control' and 'forward control'. Backward control is directed at an upstream process from that at which inspection takes place, while forward control is aimed at a downstream process. Backward control can be used where a process results in a quality characteristic which cannot be assessed immediately, perhaps not until other processes have been carried out. Forward control is used in situations where control of a process is problematic, and in which appropriate adjustment of a subsequent process allows the opportunity to rectify potential non-conformance. There is an obvious requirement for the control information to be synchronized with the product, which has implications for traceability requirements and CIM systems configuration. Figure 3.2 illustrates remote process control using the ICAM definition method zero (IDEF0) technique which is explained in Chapter 8.

Process integration involving either backward or forward control requires communication of data from one process to another, via a computer to which all relevant process control and local quality monitoring units are connected. This computer must evaluate the quality information and initiate adjustments via functional integration with the CAM sub-system, perhaps to alter process parameters for downloading within that sub-system. Taking the example of a CNC machine tool, this might involve change to a sister cutting tool, alteration to tool or work offsets, or part program parameters, depending on an analysis of the size of error detected.

Adjustment of the manufacturing process could also be achieved by functional integration with the CAM sub-system at a lower level, by direct communication with a downstream process control computer (perhaps the CNC machine tool controller) initiated from the local quality monitoring computer for the upstream process.

3.6 SUMMARY

It is not sufficient for CIM theorists to avoid the issue of quality by supposing that a general TQ concept, based on a company-wide continuous improvement process, will solve the operational problems of quality information collection and management in a largely automated manufacturing system.

The concept of IQS enables quality to take its rightful place alongside considerations of quantity in the CIM system. QC elements will form an essential feedback and control system in any viable CIM. The requirement for timely quality information available to manufacturing is clear. It must be analysed and presented for product and process control, improvement and decision support. Information is also needed by the design, process planning, production management and manufacturing functions, and throughout the business system. The effective management of this quality information

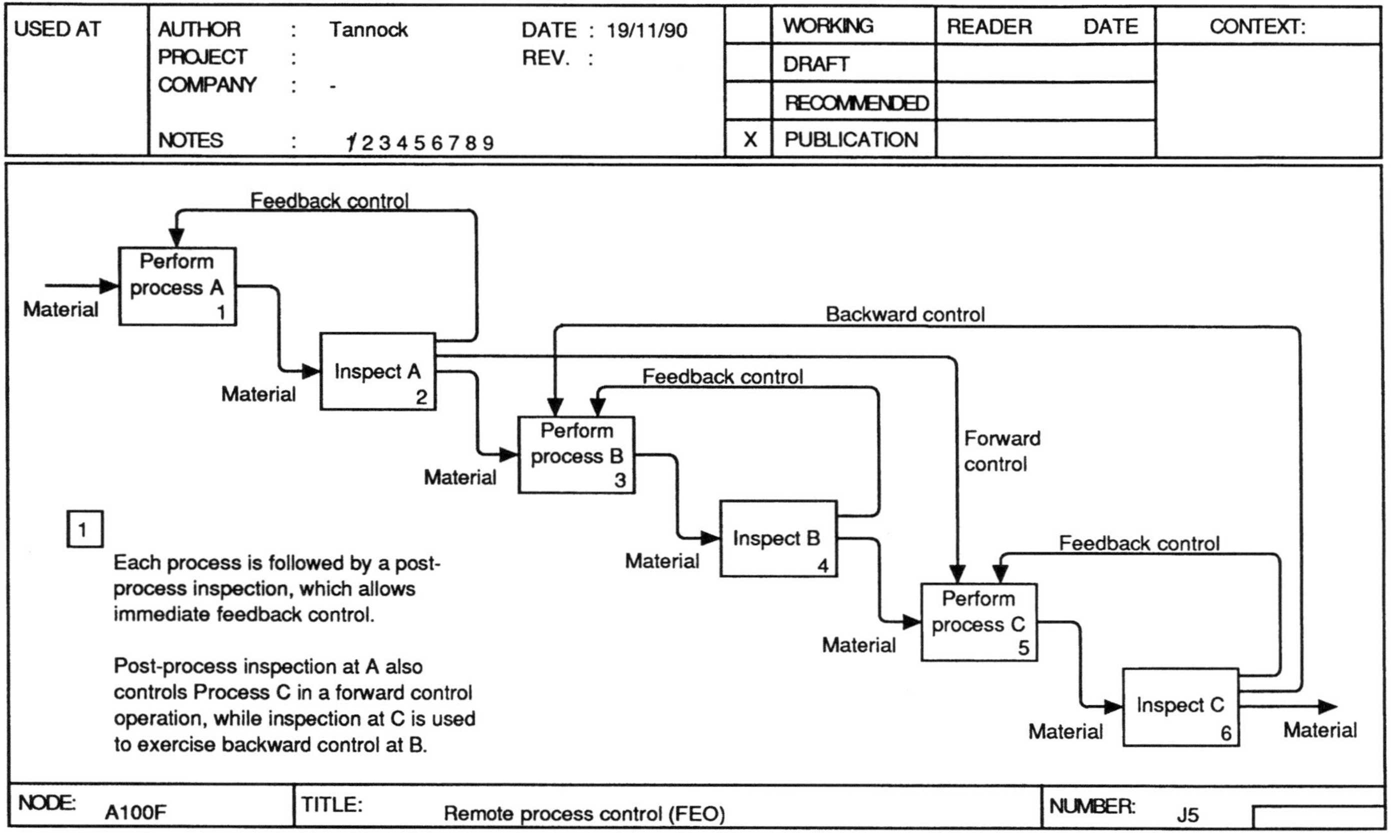

Figure 3.2 Remote process control (FEO)

requires automation and correct integration both internally and with other functions within a CIM system.

Quality systems integration must be carefully considered, and a framework for this purpose has been described, with three axes of integration: vertical (within the quality sub-system); functional (with other sub-systems); and process (through the stages of the manufacturing process).

4

Assessment of quality performance in manufacturing

4.1 INTRODUCTION

The quality performance of a manufacturing organization is a matter with which managements at all levels are frequently preoccupied. Quality performance is a complex and multifaceted matter, which many manufacturers have difficulty in fully appreciating and controlling, but which is widely recognized as among the most important factors in organizational success and survival. This chapter examines existing measures of quality performance, and proposes an alternative. Two aspects of quality performance assessment are of particular relevance to the automation of quality systems: first, the justification of investment decisions for automation, and second, the suitability of the assessment criteria for automatic analysis and presentation.

The available criteria of quality performance, listed here, will be reviewed and assessed in the following sections. They are:

1. defect levels in manufacturing;
2. value loss functions;
3. indirect measures of performance;
4. quality cost.

In the short term these measures may be used by the quality system for analysis and reporting on manufacturing quality performance, while in the longer term they could be used to assess the effectiveness of the quality system itself.

Some of the material in this chapter is rather mathematical. A certain knowledge of statistics will be required to follow the argument in some places, although the worst parts have been banished to Appendix A. The final result of the maths, however, is a simple little equation which uses batch mean and standard deviation to calculate a useful monetary quality measure from either 100% or sampling inspection data.

4.2 DEFECT LEVELS IN MANUFACTURING

Defect levels in manufacturing are a fundamentally crude measure of effectiveness, yet one which has attracted a lot of attention in recent years, especially with the popular 'zero defects' approaches to quality improvement. The criterion of defect levels in production is useful for many purposes. It is easily understandable, and requires little calculation and analysis. It may form the foundation for charts and graphs to inform and motivate the workforce in quality improvement, and provides a basic level of information for decision support.

Defect levels are usually expressed as a percentage, or sometimes a proportion, of total production. Another alternative, which can be adopted when defect levels are low, is to express them in parts per million. This gives greater impact to the figures. Defect level may be used within the quality system in order to determine the necessary allocation of resources for process improvement, or it may be used to ascertain the quality performance of the manufacturing system itself, by a comparison with other departments, plants, or over time.

There are two major disadvantages of defect level as a criterion of effectiveness. There is a lack of consequential cost information – the financial effects of the defect levels recorded. There is no formal method of distinguishing between important defects and defects resulting in negligible quality loss. In addition, there is no consideration of the costs involved in quality system operation, including the defect level assessment itself. In a practical industrial situation, the users of defect level information will, of course, form intuitive judgements or perform *ad hoc* calculations of cost as an aid to decision-making.

Despite these drawbacks, defect level data are often used in practice, because they are simple to collect, analyse and present. In addition, they are essential to the calculation of more sophisticated cost measures.

4.3 VALUE LOSS FUNCTIONS

Several value loss functions have been suggested. The basic idea is that a product's financial worth varies according to some mathematical function which is related to the degree of departure of quality characteristics (such as dimensions) from their nominal values. The traditional tolerance-based quality system used by most manufacturing companies involves an implicit 'step' value loss function, whereby product value is assumed to be lost if a quality characteristic is outside the tolerance interval. Full product value is retained even when all the characteristics are 'on the limit', in other words only just inside the tolerance band.

Perhaps the best-known explicit value loss function is that due to Taguchi (1981), who proposes that there is a 'loss to society' due to a product characteristic deviating to any extent from the nominal design characteristic. Taguchi's value loss function takes the form of a quadratic curve, and is obtained by means of certain mathematical assumptions which have not remained unchallenged, though the results have been found to be useful in many circumstances.

The notion of 'loss to society' is difficult to relate to our traditional accountancy concepts, although there is an analogy to the conventional quality cost areas described later, in that loss to a company and to the customer due to poor quality can be seen in terms of an overall loss, composed of internal and external failure elements of the total cost of quality.

The Taguchi loss function for a single quality characteristic can be represented by the following simple equation:

$$\text{Loss} = L(v) = Kv^2 \tag{4.1}$$

where the loss due to a product characteristic deviating from the nominal by v is $L(v)$; and K is a constant equal to D/T^2, D being the cost of scrapping or rework for that product and T being half the tolerance band for that characteristic.

Figure 4.1 illustrates the Taguchi value loss function. This relationship is proposed for an individual product, but might be extended to apply to a complete hypothetical batch of products which each have a single quality characteristic with a constant deviation equal to v by simply multiplying by the batch size N, hence:

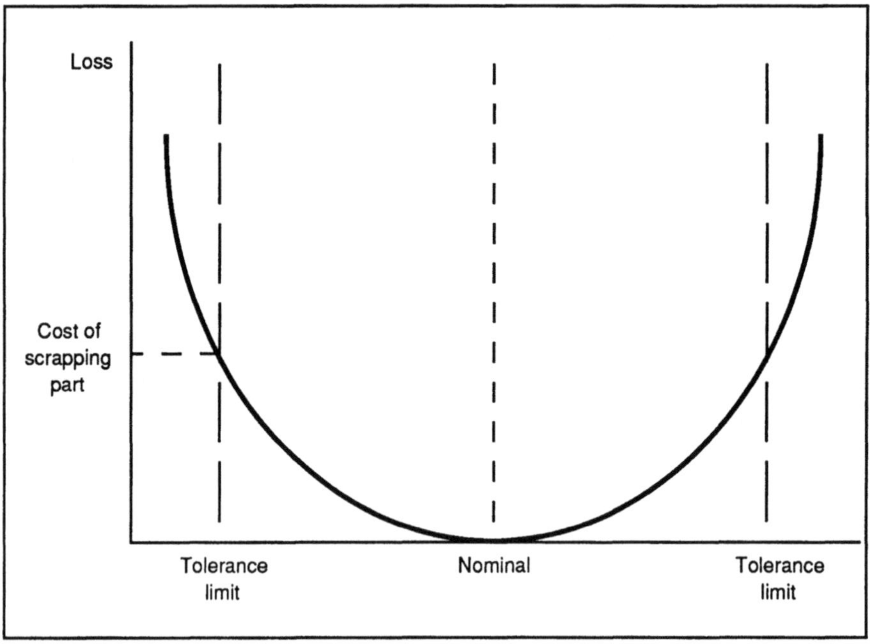

Figure 4.1 Taguchi value loss function

$$\text{Total loss} = NL(v) = NKv^2 \tag{4.2}$$

However, for a practical batch any variable quality characteristic is likely to be continuously distributed according to some distribution, so an extension of Taguchi's loss function is required for a useful assessment of quality performance. We can combine the probability density function $f(v)$ of a variable characteristic resulting from a manufacturing process subject to random variation with the loss function $L(v)$, by integrating the two functions of v together:

$$\text{Expected Loss} = \int_{-\infty}^{\infty} L(v)f(v)\,\mathrm{d}v \tag{4.3}$$

If both functions $L(v)$ and $f(v)$ are known, this would allow a numerical result for the value loss resulting from a batch of products.

There is again no indication of inspection costs or other operating costs of the quality system.

4.4 INDIRECT MEASURES OF PERFORMANCE

The usual global measures of manufacturing system performance, such as productivity and profitability, are based on accounting concepts. It is very difficult to relate the influence of each variable (for example, direct labour productivity or stock turnover) in a manufacturing system to its individual effect on global measures. Only in extreme cases of exceptional quality performance (either good or bad) does the quality factor become pre-eminent and dictate the overall manufacturing system performance. Normally it might be expected that the causal relationship is reversed, and the overall system performance (in terms of work organization, structure, culture and skills) determines the manufacturing quality performance.

Ignoring (at our peril) such intangible factors, and examining only the quantitative factors, such as stock levels and production efficiencies, is of no great help. There is at present no complete mathematical model of manufacturing performance able to evaluate the effect of all the quantitative factors, including quality, in the general manufacturing situation.

A number of companies use customer complaint levels, service calls or warranty claims to indicate the quality performance of the organization. These measures seem superficially attractive, because at first sight they would appear to be directly related to customer satisfaction – normally the quality goal of the organization. However, in practice, the type of customer and sector of industry may be more important than the quality of the product. Customer reactions to quality performance will be affected by a multitude of factors connected with attitude, expectations and the changing competitive position with regard to both price and quality. In addition, customers often have insufficient knowledge or information to assess quality performance effectively. Customer satisfaction may be more effectively measured by surveys of recent customers, or investigations of the level of repeat sales.

4.5 QUALITY COST

The best-established criterion of effectiveness for the assessment of quality performance is quality cost. The quality cost concept embraces three or four elements:

1. appraisal costs – the costs involved in inspecting and monitoring the quality of manufacture
2. prevention costs – the costs of action taken to investigate and prevent defects in production processes, quality planning, training and the introduction of improvement programmes
3. defect (internal failure) costs – the internal costs of failing to achieve quality standards, including disruption, concessions, scrap and rework costs;
4. failure (external failure) costs – the costs incurred due to failure of products after they are received by the customer, including the consequential losses of reputation, goodwill and future business.

where internal and external failure costs can sometimes coincide.

In the traditional analysis, these various costs are totalled to give the overall cost of quality (COQ), a concept usually attributed to Juran (1979). The widely accepted theoretical model representing the relationship of cost to quality level, showing an optimum (lowest) quality cost, is shown in Figure 4.2. This relationship is difficult to validate in practice, as the quality level of production is not an independent variable with which it is easy to experiment, and very few companies will have such data available. Alexander (1983) reports that an attempt was made to demonstrate the validity of the COQ model using quality cost data from a Marconi Avionics plant. The attempt failed: despite the collection of a large amount of data, the researchers were unable to confirm a relationship of the form illustrated, or to find any other consistent pattern.

Schneiderman (1986) and Bajpai and Willey (1989) have suggested that a different relationship between prevention cost and quality level might be expected at very low defect levels, where Juran proposes that prevention costs must be driven rapidly to high levels. This means that the total quality cost is not at an optimum when quality is perfect – a conceptual problem which has long troubled the advocates of zero defects approaches. Bajpai and Willey (1989) envisage the relationship illustrated in Figure 4.3.

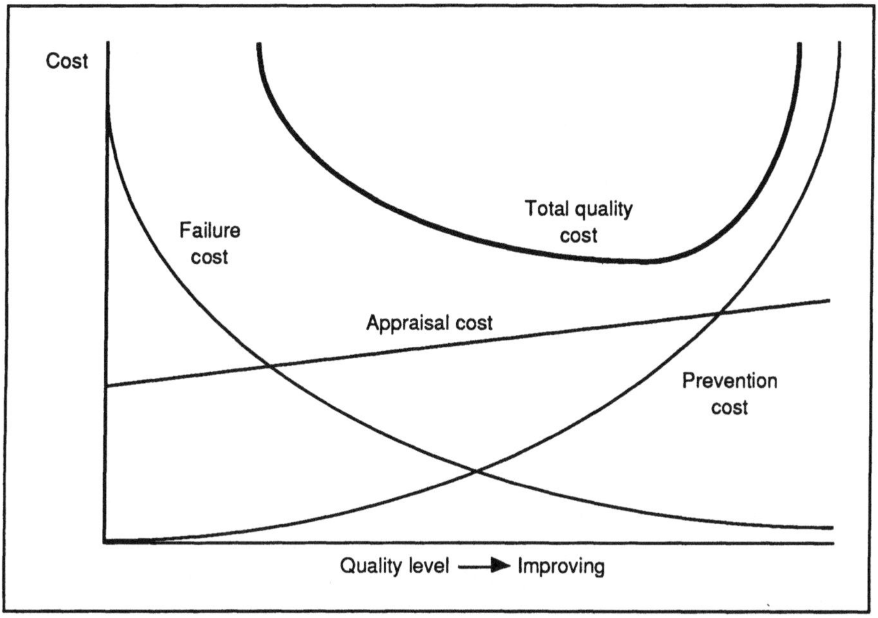

Figure 4.2 The Juran quality cost model

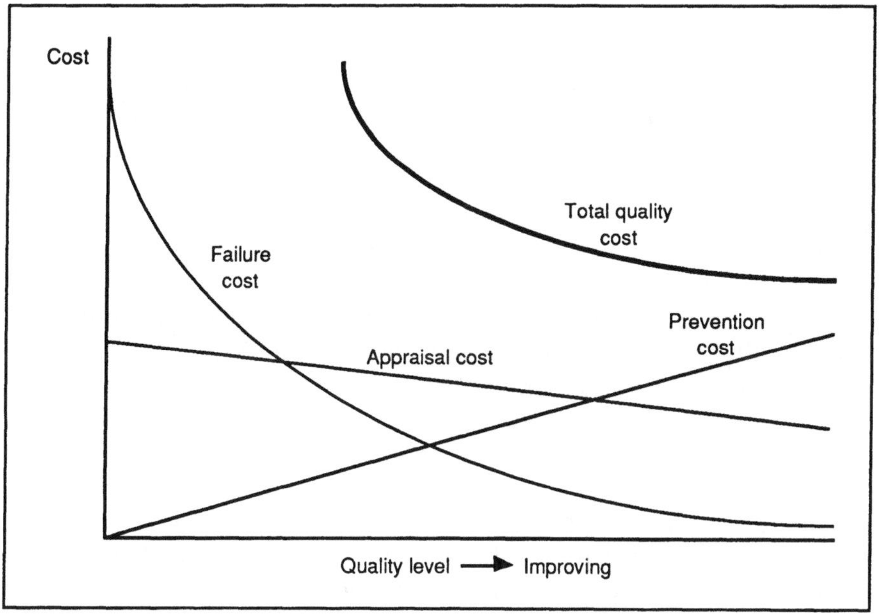

Figure 4.3 Quality cost model according to Bajpai and Willey

It must be emphasized that the costs are represented on these graphs by curves which have not been derived from mathematical analysis (or, as noted above, confirmed in practice). This does not stop the COQ approach being included in most textbooks, and taught to most students, if not implemented by all manufacturing companies. The sole basis for this continued popularity of the quality cost model might be summarized as 'well – it seems sensible'.

Accepting for the moment that the COQ approach is viable, the financial aim of quality systems where quality costs are collected is to minimize the costs incurred in the internal and external failure categories, on the assumption that they are greater than the sum of the other costs. This is considered to be true in the 'zone of improvement', where the quality level is below the optimum (minimum) total quality cost point.

Management can make decisions to substitute voluntary quality costs for involuntary costs, by introducing prevention and improvement programmes. These prevention and improvement activities should be seen, according to modern TQM thinking, not as a cost but as a vital investment. The importance of this strategy has been emphasized by a number of quality gurus, such as Crosby (1979), who makes the point in the title of his influential book, *Quality is Free*. Crosby refers to the COQ in the following terms:

> Used as a management tool for the purpose of focusing attention on quality management the COQ is a positive blessing and serves a unique purpose. Used as an accounting measurement, like the calculation of nuts-and-bolts inventory, it becomes a useless pain.

Various national standards exist on the determination and use of quality related costs, for example the British Standard BS 6143, which is under revision. Part 2 of BS 6143 was issued in 1990 to replace the content of BS 6143:1981, and Part 1 is to follow. This part will deal with quality costs in terms of the costs of conformance and non-conformance, with particular emphasis placed on process and service.

Such standards have been criticized, on the basis that the cost elements listed in the standard do not reflect the incidence and distribution of costs in manufacturing, and consequently there are difficulties and inconsistencies in the categorizing of quality costs. In addition, the resource implications of the quality cost collection and analysis exercise are often considered a practical disadvantage.

The truth is that quality costs are not used very extensively in industry (perhaps by only one-third of companies), because an appreciation of their uses and the advantages of their collection is not widespread. The analysis of quality costs is usually separate – where it is attempted – from the management accounting systems. Quality costs are not assessed effectively by conventional accounting systems; and many costs associated with quality activities find their way into manufacturing or general overheads. Professional accountants working in manufacturing industry often have little awareness of basic concepts in quality economics, and hence find difficulty in the assessment of quality performance.

There is little doubt that the COQ approach to the assessment of quality performance is in trouble. Lacking a theoretical basis, troubled by internal conflicts, unconfirmed by experimental evidence, burdened with inadequate standards and little used, the whole COQ approach might be said to be in disrepute. However, there are some valuable ideas which might be salvaged. In the rest of this section we start from the basics, to examine what sort of decisions we could make on the basis of a simple quality cost approach in a batch manufacturing situation.

In the short term, where existing inspection staff and equipment are used, the appraisal, internal and external failure costs for a batch may be reasonably expected to be straight lines with respect to defect level. The prevention cost, however, is of a different type to the other costs. If prevention spending is to be considered as part of a company-wide policy of investment in never-ending improvement, it does not seem appropriate to consider the prevention costs as part of a simple model. Hence the 'cost of quality' analysis given here does not include prevention spending.

A break-even analysis of the costs of different inspection strategies may be performed as follows. Let C be 'cost of quality' for the batch; N the batch size; P the proportion defective in the batch; D the cost of scrapping/rectification per article; F the cost of defective article reaching customer; and A the cost of inspection per article. Then the cost of no inspection is given by:

$$C = NFP \tag{4.4}$$

while cost of 100% inspection (assuming no inspection errors) is given by:

$$C = NA + NDP \tag{4.5}$$

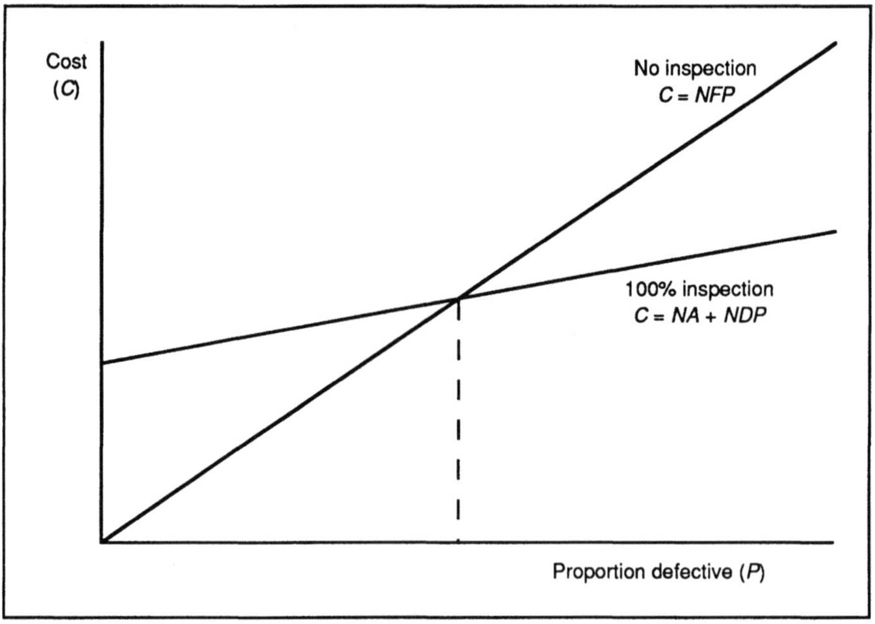

Figure 4.4 A simple model of batch quality costs

Figure 4.4 shows a generalized break-even diagram for these two inspection strategies. It will be seen that the type of inspection strategy giving the lowest total cost is determined by the proportion defective in the batch. Analyses of the cost of sampling inspection may also be performed, if information is available regarding the operating characteristic of the sampling scheme used. We could hope to make some simple decisions on the basis of information provided by this type of analysis. It would be possible, for example, to decide on the type and level of inspection which would be appropriate for each quality characteristic, or justify investment in inspection automation.

4.6 AN ALTERNATIVE APPROACH

From this review of the existing criteria of quality performance, it can be concluded immediately that defect level, with its lack of cost information, is too crude a measure to provide a really effective guide for decision-making on quality improvement. Furthermore, indirect measures of performance are not yet sufficiently well developed to allow their use as effective criteria of quality performance.

The value loss function as a measure of manufacturing and quality systems performance is of greater interest. The influential Taguchi approach emphasizes continuous reduction in variability, rather than the use of tolerances; this is considered to be aligned with a customer-centred approach. However, the cost of operating the quality system is not taken into account, making the approach unsuitable for investment justification.

Quality cost (COQ) is the most widely used of the existing measures. The approach has been thoroughly criticized above. However, quality cost is the only currently accepted criterion of effectiveness which may be applied to the complete quality system. The approach which will now be proposed as an alternative is one which combines the appraisal cost element from the traditional total quality cost analysis with a value loss function.

The Taguchi approach defines a mathematical relationship between distance from nominal, and a financial loss which we may choose to equate with the combined internal and external failure cost elements of the COQ. As an example, considering 100% inspection, replacing the appropriate term in equation (4.5) by the value loss term from equation (4.2) gives the cost of quality for a batch as the sum of all the losses due to the individual deviations from nominal, added to the appraisal cost:

$$C = NA + \sum_{i=1}^{N} K v_i^2 \tag{4.6}$$

For sampling inspection, where n products are sampled from the batch, we can obtain an estimate of the parameters – the mean (μ) and variance (σ^2) – of a continuous distribution of the deviation (v) from nominal for some variable quality characteristic. Taking into account both the Taguchi loss function and the distribution of this characteristic by means of its probability density function $f(v)$, we can show that:

$$C = nA + NK(\mu^2 + \sigma^2) \tag{4.7}$$

(see Appendix A for derivation of this equation). The quality measure C can be easily calculated on paper or by a simple computer program, as shown in the example below.

It should be noted that there is no requirement for the continuous distribution of errors from nominal to be a normal (or any other specified) distribution. It must also be emphasized that cost information of this sort is a comparative quality performance indicator, and must not be confused with the use of variable sampling techniques for acceptance or rejection, although the same data might be used for both purposes.

Because sampling inspection is used, μ and σ^2 in equation (4.7) will only be estimates of the true statistics of the entire batch. The sample mean is the best unbiased estimator of the batch mean. If the sample variance is calculated from measured data using $n - 1$ degrees of freedom for the estimate – that is, the denominator in the calculation is $n - 1$ – then the sample variance obtained will also be the best unbiased estimator of the batch variance. If specified confidence intervals for the estimates of μ and σ^2 are required, the necessary sample sizes may be readily calculated.

In the following example, suppose a batch of 500 components is manufactured from which a sample of size 25 is inspected at a cost of £1 per component. In addition to the usual acceptance information obtained from this inspection, one or more vital characteristics are selected and the mean error from nominal (μ) and standard deviation of the error (σ) calculated, as if for a process capability study. The selection of the critical characteristics might be performed using the same type of criteria as would be used for the selection of the feature to chart using SPC techniques.

In this case we will take the figures obtained from the measurement of our chosen variable characteristic (a dimension) to be:

mean error (from nominal)	$\mu = 0.07$ mm
standard deviation of errors	$\sigma = 0.16$ mm

The loss due to a defective component will be assumed to be £10. The tolerance band has a width of 1.0 mm. Hence our factor $K = 10/(0.5)^2 = 40$.

The quality performance measure C may then be calculated using equation (4.7).

$$C = (25 \times 1) + (500 \times 40) \times [(0.07)^2 + (0.16)^2]$$

$$= 25 + 610 = £635$$

If the mean deviation from nominal is improved to 0.02 mm and the standard deviation to 0.05 mm by better control of the manufacturing process, C falls to £83, showing the sensitivity of the quadratic loss function to the quality improvement.

4.7 CONCLUSION

The simple composite measures of effectiveness suggested above are easily computed, using statistics which may be already available in a company using SPC. The quality performance is measured in monetary units, which is often considered to be an advantage when presenting figures to senior management.

If used to assess the costs of operation of the quality system, these measures will concentrate attention on the continuous reduction of product and process variability aimed at the nominal characteristic. They will also allow the financial benefits of reduced appraisal costs, whether arising from inspection automation or reduced inspection requirements due to improvement of the manufacturing process, to be assessed.

The quality performance measure C is also well suited for use with an automated quality system incorporating a central database. It is not necessary to collect a vast amount of complex quality cost data to achieve a meaningful figure. A selection of critical product features could be agreed, and the aggregate result automatically calculated and reported periodically as an indicator of the success of the improvement.

Part Two

Quality systems – design and innovation

In Part One the background of quality thinking and the rationale for quality systems automation were explained. For any other aspect of manufacturing it might now be considered appropriate to launch directly into the technology of automation – the instruments and computers which are available to carry out the functional activities required.

Quality is a rather special area, however. There are important reasons why quality activities have not been automated to the extent found in other fields of manufacturing, such as production operations, scheduling and management. These reasons are connected with the unique double aspect of quality: it is both a functional area concerned with routine information management and a business strategy for excellence. Automation of quality functional activities must have implications for a company culture aimed at total quality.

To reconcile these two aspects of quality is not easy. In the search for an integrating theme, the importance of quality information for improvement has already been emphasized. Part Two goes further, and takes a conscious systems approach to quality. Quality practitioners tend to see 'quality systems' mainly in terms of the rather dry and prescriptive edifice of organization and procedures laid down in quality systems standards like the ISO 9000 series. But systems science is a branch of knowledge which has far more to offer than is often supposed. Using a systems approach it is possible to attempt a reconciliation between the different aspects of quality, by incorporating the organizational, human factors and functional aspects of quality into a single framework.

The first chapter in this part reviews some of the systems approaches which are valuable in the quality area. Next, innovation in quality systems is explored, and a planned evolution process proposed which provides a practical structure for the major changes which will be required to introduce automated systems.

Quality systems design and improvement is tackled with the aid of a powerful structured modelling technique which can incorporate elements as diverse as 'quality enthusiasm' and 'batch quality data file' into a unified systems description. Finally an information technology strategy for quality systems automation is proposed, incorporating a configuration for automated quality data collection and management capable of integration with computer-integrated manufacture.

5

Systems in quality

5.1 INTRODUCTION

An important theme of this book is the conscious application of a systems approach to quality. Systems thinking in technological and management disciplines is now considered to be vital to effective design, understanding and problem-solving applied to complex business functions. On occasion, however, the word 'systems' can be little more than fashionable jargon, and the use of the word should not be taken as confirmation that any conscious systematic thought has occurred.

Systems thinking encompasses a body of theory, comprising concerns, concepts and techniques, which can be applied to QA activities, just as to many other human undertakings. Checkland (1984) states that: 'The core concerns are the two pairs of ideas: emergence and hierarchy, communication and control.' The hierarchy is that of components, sub-systems, and systems leading to the emergence of system characteristics at certain levels of complexity, which cannot be explained from an analysis of component parts of that system. The control of systems is exercised by the analysis and design of measurement and communication processes which integrate the components of the system.

Any system requires an inherent rationality, in which is reflected the arrangement, linking and control of its elements in order to achieve specific objectives. These objectives are derived from the underlying values and principles on which the system is based. Almost every system will form a sub-system of a larger system, and the principles on which it is based may be seen, in the context of the larger system, in terms of specific objectives.

Systems concepts and techniques are numerous, but all are concerned with the need to provide a holistic approach to problem-solving and decision-making, related to systems design, analysis and the description of complexity.

The relevance of these matters to the achievement of quality in manufacture cannot be over-stated. QC and QA are an important area of application for systems thinking. This chapter attempts to examine how systems thinking can be applied to quality. In order to do this, the ways in which systematic approaches currently influence quality activities will first be reviewed.

5.2 QUALITY SYSTEMS

A quality system in manufacturing can be seen in systems terms to be a human activity system, rather than a natural, physical or abstract system. The quality system may also be seen as a sub-system of larger human activity systems, such as the manufacturing system, and the total organization, and related to other human activity systems in different organizations. Within the quality system will exist sub-systems which can be physical in nature (an automatic inspection sub-system, for example). In addition, there are abstract systems which may be closely related or even partly coincident with individual quality systems, such as a philosophy for TQM, or an international standard for quality systems. Close relationships exist with other sub-systems of the organization, particularly design, logistics, manufacturing and production control.

Most managers and staff in industry and commerce are influenced, consciously or unconsciously, by systems concepts which should guide their thinking towards effective achievement of objectives. A perception of the business function, organization and commercial environment in a holistic manner (as systems composed of component sub-systems, with mutual connections which influence each other in complex ways) is nowadays to be expected. Those in quality assurance functions are likely to be exposed to systems ideas because of the nature of their roles, and through the holistic themes of TQM. An understanding of systems concepts should lead managers to make better decisions, based on the requirements of the organization as a whole rather than local needs, thus providing the framework for greater rationality and efficiency in management.

There has been considerable application of systematic thought in quality, and the term 'quality systems' is used to describe the results. ISO 9000 defines this term as 'the organizational structure, responsibilities, procedures, processes and resources for implementating quality management'. The elements which are specified in the ISO 9000 series of standards relate to:

1. management responsibility;
2. contract review;
3. quality system;
4. design control;
5. document control;
6. purchasing;
7. purchaser-supplied product;
8. product identification and traceability;
9. process control;
10. inspection and testing;
11. inspection, measuring and test equipment;
12. inspection and test status;
13. control of non-conformance;
14. corrective action;
15. handling, storage, packaging and delivery;
16. quality records;
17. internal quality audits;
18. training;
19. servicing;
20. statistical techniques.

It might be expected that a quality systems approach would also stress the effective analysis and communication of quality-related information to allow management decision support and hence improved business control, and emphasize the relationship with other business sub-systems. However, recent national and international standards for quality systems give little consideration to communication or integration. There are sections relating to the reporting of audit findings, but quality information is mainly viewed in terms of records to verify the achievement of the required quality, rather than as a source of information for improvement or decision support. This is a serious failing in systems terms, because it inhibits the full integration of quality functional activities into the organization, and fails to take a comprehensive view of quality as it relates to the entire business.

Quality systems standards seem to stem from rational management thought, rather than from conscious systems thinking. They do, however, represent a systematic approach. In particular, the clear specification of terms, policy, objectives and procedures is important to the application of systems thought. The QM approach usually adopted combines accepted management methods with the application of a quality systems standard, often linked (sometimes in a rather uneasy alliance) with one of the various TQ philosophies.

In restricting the scope of quality systems to the areas covered by the national and equivalent international standards, we ignore some of the major contributions which a comprehensive systems approach can bring to quality. There are important further areas of activity to which systems techniques may be applied. These are design, improvement and innovation related to quality systems.

These activities are closely linked, as most practical situations will involve aspects of each. They are fundamental activities in the process of improvement which is the key to success in manufacturing, encompassing large-scale innovatory changes as well as *kaizen* (a Japanese word which represents a continuous process of – often relatively inexpensive – small-scale improvement).

Design and innovation for quality systems will be the subject of later chapters. First, however, some systems ideas which might be useful for quality systems will be reviewed, with particular reference to their use in quality problem-solving. It should be appreciated that, despite the ongoing attempts to develop an all embracing general systems theory, systems science still comprises a number of disparate approaches. A widely accepted division is that between 'hard' and 'soft' methodologies, which will both be reviewed here. In addition, a section on computer systems analysis and software development methods is included. Although computer systems analysis is not considered to be in the mainstream of systems thinking, it involves the use of systematic methods and techniques to deal with complexity and solve specific design problems – thus surely qualifying for consideration here.

5.3 RATIONAL SYSTEMS – THE HARD SYSTEMS APPROACH

The 'hard systems' approach is based on the idea that a situation or problem may be sufficiently well understood for its present state to be defined, alternatives generated and then a desired state or optimal solution selected in some rational way – for example, by economic analysis of the costs of the alternative strategies. Hard systems approaches are represented by systems engineering and systems analysis. The latter should not be confused with computer systems analysis, which, as will be seen later, is a field distinguished by the use of certain techniques for the specification, design and development of computer systems and software.

Hard techniques are based on **positivism**, the rational philosophical background to the objective methods of the natural sciences. As a result of its rational positivist basis, typical hard systems thinking tends to ignore the human factors that often have a major part to play in the system operation and in the decision-making process. Underlying assumptions are made about the nature of man, who is considered to be deterministic in nature, much like a complex machine.

The rejection of this deterministic view of man in all TQM thinking is clear. For a TQM approach to be effective, the very qualities ignored by a mechanistic, 'theory X', view of man must be fostered. Particularly important are ideas connected with the working group, where the whole effect is conceived to be greater than the sum of the individuals who are the components of the group. Hard systems approaches, therefore, tend to be in conflict with the human factors themes of TQM.

Paradoxically, with the advent of computer solutions to manufacturing systems requirements, the problems of complex system representation and design have acquired a new urgency, with a requirement for the efficient integration of computer systems into CIM. It is widely recognized that some formal system design methodology is essential for the design of CIM systems, and desirable for other complex systems incorporating elements of computer integration. The lack of a suitable design methodology is held to lead to 'islands of automation' where incompatible computer (and manual) systems form a barrier to further systems development. The need for effective design and appraisal tools for complex systems has led to an increased use of hard systems techniques as well as computer systems analysis methods in manufacturing.

Closely associated with the hard systems approach are the methods of operational research (OR) or operations management analysis. Much of the academic literature on quality matters is based in OR, where reductionist methods are used to create a mathematical model of a real-world situation and draw some conclusions about the real world on which, it is hoped, engineering and management decisions can be based. Applications of OR techniques in the quality area are found in optimization of sampling schemes, in inspection station placement and in process response and control models.

The disadvantage of work of this type stems from the OR methods themselves. To provide results, the models must be stripped of the realistic details and complex influences which are the important characteristics of any real-world problem. Only a small proportion of work of this type has immediate practical application in quality assurance, although it may set out possible themes for future developments. The works of Shewhart (1931) with statistical methods, and Dodge and Romig (1941) using probability, are the progenitors of this theme in the quality field, and their influence cannot be denied. Statistical and probability tools and techniques are now very widely used for the analysis of technical data, and hence for control and improvement. The use of statistical tools such as process capability studies and control charts is widely supported by the proponents of TQM. However, these methods, like more recent developments in OR, do not have the scope to tackle broader system design issues.

5.4 THE SOFT SYSTEMS APPROACH

In the search for an approach which takes into account the human factors themes so important in TQM, recourse must be made to soft systems thinking. In this approach, the subjective perceptions of individuals and social or work groups are considered to be of importance. There is no single testable account of a human activity system, but a number of individual accounts according to the position and perspective of each 'actor'. Soft methods tend to emphasize hierarchy and emergence, rather than communication, measurement and control. They are distinguished by the use of techniques which attempt to maintain the 'richness' or complexity of the systems description, unlike hard techniques which use a reductionist approach to refine a situation down to its essentials for analysis.

Soft systems methods tend to be perceptive and insightful rather than objective. While the importance of this approach cannot be denied, its lack of objectivity is a disadvantage for many practical purposes, and particularly for the design and description of quality systems, for which objective standards exist.

There are formalized analysis and design methods, such as the soft systems methodology of Checkland (1984), and Beer's viable systems model (described by Jackson, 1988) but such approaches tend to be conceptually complex and require considerable training and experience for effective use.

Hard and soft systems approaches differ in their definition and treatment of problems. For the hard approach, a problem is some situation in which there exist, according to Flood (1988):

1. decision-maker(s);
2. objectives which may be defined;
3. alternative means (solutions) of achieving these objectives.

The soft approach characterizes the problematic situation as one perceived to be deficient in some way, but which cannot necessarily be formulated in terms of a search for an efficient means of achieving a desired objective. To do so might lose the richness and complexity which contribute largely to the perception of the situation. The definition of purposes and goals is itself, therefore, a problem. The search for solutions is replaced by the search for improvement, recognized by an overall reduction in dissatisfaction. To quote Checkland (1984) again: 'the unquestioned prime value embodied in "a systems approach" is that continuous, never ending learning is a good thing'. The analogy with the TQM theme of never-ending improvement is striking.

Many quality problems can at first appear relatively well defined, because of quality system requirements for formal specifications, procedures and techniques, which have been developed to reduce the uncertainty in a problem situation. Formal and informal aspects of quality function activities may be distinguished. The formal aspects of a quality problem are likely to be described as being concerned with technical difficulties (such as non-conformance to specification) or difficulties concerned with information (such as a lack of correct quality data).

Quality problems which might appear technical in nature may actually be concerned with the functioning of the quality system itself, and with the methods by which the formal procedures are defined and applied. The clear definition of a problem, from one point of view (that of the manifest quality specification or procedure) does not necessarily assist the generation of a solution. The problem may indeed consist of a conflict between different views of the specification. Real-life quality problems tend to be highly complex, involving informal elements such as differing perceptions, poor communications, inter- and intra-company relationships, history and politics. These may overshadow the formal cause of the problem and restrict the ability of the people involved to resolve it satisfactorily.

As an additional element, quality problem-solving may be seen as part of an institutionalized process of continuous improvement. There may be special organizational mechanisms, such as quality circles or continuous improvement teams (CITs), designed to ensure that the workforce is effectively involved in problem definition and solution.

Any choice of problem-solving method will be influenced by the systems approach adopted, whether this is explicit or implicit. There seems to be a fundamental conflict between the hard systems techniques (with which are associated many commonly used techniques for problem definition and solution), and the soft human factors themes central to TQM approaches. Unfortunately, as suggested earlier, the softer systems analysis techniques, such as the soft systems methodology, are conceptually complex and difficult to implement. To solve real quality problems, individuals and teams need tools and techniques which are easy to learn and apply.

It seems that a pragmatic approach to problem-solving, making use of hard problem-solving techniques for technical problems and of soft systems-based techniques for organizational and human issues, is most appropriate. Using such an approach, the problem-solver is a practitioner with a toolkit of techniques selected and used in a heuristic fashion.

5.5 COMPUTER SYSTEMS ANALYSIS AND SOFTWARE DEVELOPMENT

The design and implementation of large software systems, such as CIM systems, is one of the most complex tasks currently undertaken, and special methods and approaches have been developed to cope with this complexity. Assurance of quality in software development is a key activity, where many of the ideas of product QA are combined with concepts peculiar to software development, together with a great emphasis on full documentation. The formal software development activity is split up into phases, which make up the software life cycle. These are:

1. requirements analysis;
2. system specification;
3. system design;

4. coding (or programming);
5. integration;
6. operation;
7. maintenance.

The starting point is the customer's or user's statement of needs for the proposed system. As customers are rarely computer experts, this is usually naive, ambiguous and incomplete. Hence the first two phases, requirements analysis and system specification, consist of techniques for the refinement and formal expression of these needs in the form of functional and non-functional requirements. These early phases are critical to the success of the project, as a poor systems specification will lead to cost and time overruns later on. The systems requirements are usually expressed in everyday (or 'natural') language, although there is now a tendency to use formal mathematical notations which are more exact, but more difficult to use.

The design phase consists of systems design, where the overall systems architecture is defined to satisfy the systems specification, and detailed design, where the systems design is transformed into a form with which a programmer can work. The latter involves the design of individual algorithms, or rules, to perform the required functions. Special program design languages are often used to define the required algorithms. There is a tendency to automate the production of program code from the detailed design.

The coding or programming phase, which most laymen imagine is the key activity, can indeed be virtually automatically performed if systems design has been properly carried out. The software is developed in the form of modules or units, each of which implements a particular algorithm, and which are sufficiently limited in size and scope to be amenable to effective testing and validation. Whether or not it is manually performed, this phase employs yet another language, probably one of the high-level computer languages such as C, FORTRAN, COBOL or ADA.

In the integration phase, the modules are linked together into the final application, one at a time to ensure that the testing program is able to identify most easily any faults in each module. The final product is tested again to verify that it satisfies the original customer requirements.

After the software has been put into operation, it is almost certain that further problems will come to light. At this stage the maintenance phase is entered, and the faults are rectified. Maintenance also covers the changes which result from changes in the environment in which the software operates, resulting from new procedures, computer hardware or revisions of other software.

The entire software development cycle is often carried out using a particular (sometimes proprietary) methodology, which incorporates techniques suitable for each phase. The methodology is meant to allow the full complexity of the project to be broken down into manageable portions to which these techniques may be applied. Nevertheless, the management of a large software development project is a very demanding activity, and many efficient and respected companies have failed disastrously to meet their customers' requirements in time and to budget.

Some customers, particularly the government and military, insist on the use of a particular methodology and specify every aspect, from the documentation procedures to the programming languages used. One such methodology is the Structured System Analysis and Design Method (SSADM) which has been adopted by the British government.

A lot of effort has been expended in the development of software tools to support software specification, design, automatic coding and project management. These computer aids are known as computer-aided software engineering (CASE) tools and include analysts' workbenches and integrated project support environments (IPSEs).

In conclusion, the main theme of computer systems analysis is dealing effectively with systems complexity. Although user requirements are the starting and finishing points of the software development cycle, the majority of the work is undertaken by software experts who may have little understanding of the functional area, such as quality, with which they are dealing. These experts are often working with methodologies, techniques and tools which are themselves complex, and sometimes only marginally adequate for the task. It is not surprising that the software development process, even when fully and competently carried out, does not always result in successful applications.

As regards problem-solving, one of the fundamental systems science activities, it is clear that computer systems analysis methods are little concerned with solving everyday problems. Their considerable power is aimed at one goal – the problem of providing a computer solution to satisfy the customer.

5.6 A SUITABLE SYSTEMS APPROACH FOR QUALITY

Hard systems approaches provide many useful techniques, but these are based on deterministic concepts which deny the importance of human factors such as creativity and teamwork, and are in conflict with the ideas underlying a TQM philosophy. Soft systems approaches are based on much more acceptable concepts, but at present do not offer simple and usable techniques for practical design, innovation and problem-solving in quality.

Where are we to find a compromise approach which incorporates powerful practical abilities without denying the human factors? One answer can be found in the encouragement of a process of mutual learning and understanding between all those concerned with the system. This is at the basis of all practical quality improvement activities, and will be fostered by systems techniques which incorporate a process of dialogue. Through this process, system users can arrive at an agreed description of the existing system, then formulate objectives and methods to achieve these objectives. It implies an evolution process which promotes progress from the existing situation towards a better future system.

To achieve this progress for real quality systems, a technique must involve both a process of mutual learning and agreement, and a final outcome which constitutes a useful description and analysis of the system. In order to be valuable for automated systems design and improvement, it must be capable of representing systems complexity in an understandable manner, and must also be capable of forming the basis for a software specification. Fortunately, a technique which has these advantages is available, with the rather unwieldy title of ICAM Definition Method Zero or IDEF0. IDEF0 is used to produce systems descriptions or models, and has been widely used for manufacturing systems design in recent years. The use of IDEF0 is described in Chapter 7.

5.7 SUMMARY

In this chapter, the concept of 'quality systems' has been examined from a systems science viewpoint. Current quality systems standards are described as lacking in some of the aspects necessary for an effective systems approach. Two areas of interest in quality activities, where a systems approach should be useful, have been identified. These are quality systems innovation and the systematic design and improvement of quality systems.

Three very different systems approaches – hard systems, soft systems and the specialized methods of computer systems analysis – have been outlined and assessed from the viewpoint of their application to quality systems and quality problem-solving.

An appropriate systems approach to support automation in quality must strike a balance among the hard systems methods which are useful for optimization, computer systems analysis techniques which form the basis for software specification, and the soft systems approach incorporating a strong human factors element, which is aligned with TQM thinking.

The conclusion is that none of the currently available systems approaches meets the needs of the quality area completely, but that a mixture of techniques from hard and soft systems can be valuable.

Any serious attempt to automate quality systems must involve the use of a computer systems analysis and software development approach, but it should be based on a very clear idea of the systems required, particularly with respect to the implications for a TQM culture. This is too important to be left to the computer systems analyst, so the basic systems design should be carried out by the quality experts, using appropriate systems modelling tools.

For the systems design and improvement task a powerful systems technique known as IDEF0 exists, which offers the opportunity to describe systems in a way useful to the formal quality systems, software development and human factors aspects.

6

Innovation in quality systems

6.1 INTRODUCTION

It is important to make the distinction between innovation and other types of change. Not all change is for the better. Systems, like virtually everything else in life, degrade and lose their effectiveness unless an effort is made to maintain and improve them. Anyone who has experience of a quality system will realize how readily important procedures can become neglected and disregarded under the pressure of immediate events.

As regards change, there are three possible states which a quality (or other) system may assume:

1. Operation, where the status quo is deliberately preserved and no change occurs. This could perhaps be the case with a quality system which has been designed (perhaps by an outside consultant) in accordance with a standard, and documented in a quality manual which is then treated as if written on tablets of stone.
2. Continuous improvement, where a deliberate effort is made to assess the effectiveness of system operation, and make enhancements to correct the deficiencies which become apparent. This type of approach is usually based on an audit process, and results in relatively minor incremental changes which need little in the way of investment or training.
3. Innovation, which will result in a step change in function for the system. The introduction of quality systems automation on a large scale is an example of innovatory change. The resource requirements and resulting effects are quite different from those of a continuous improvement process.

The author was once privileged to be present at a lecture delivered by W. Edwards Deming, during which the venerable guru illustrated the distinction between an improvement process and innovatory change with the aid of one of his characteristic 'normal distribution' diagrams. Commercial performance, he explained, is a variable which is usually dependent on a large number of unrelated and individually insignificant factors, and thus we might expect the effectiveness of firms to be normally distributed, with the majority close to an average performance. He drew a bell-shaped normal curve on the blackboard. The organization which adopted an effective improvement process would influence these factors, and move its performance along the axis to the right-hand tail of the distribution. Here were to be found the most successful firms which relied on improvement alone. Then he drew a bold line far to the right of the curve. Here was the company which had something unique. A 'special cause' was at work, and the effect was a step change – a huge improvement – in performance. It might be the consequence of organizational, systems, technology or product innovation. This was the only way, he proposed, that Western companies could hope to tackle competition from Japan and other Pacific rim manufacturers.

6.2 THE MANAGEMENT OF INNOVATION

As Deming suggested, different varieties of innovation may be distinguished, although in practice most projects encompass aspects of more than one type. For our purposes, the most important are:

1. information technology innovation;
2. organizational innovation;
3. manufacturing technology innovation;
4. product innovation.

Quality systems innovation is mainly concerned with the first two of these, although we must appreciate the implications of manufacturing technology and product innovation for the quality function. When proposing technical innovation, we should take particular account of the organizational and human factors involved.

Automating quality systems might initially appear as a convergent problem in engineering and computing, concerned with the selection or development of appropriate hardware and software. Experience will soon show, however, that this approach is unable to accommodate the human factors themes in quality. These are of such strength and importance as to seriously impede naive attempts to propose automation and integration through the use of computers. It has many times been demonstrated that the 'best' technical solution is not necessarily relevant.

The need to develop a workable synthesis between the human factors themes of TQM and practical considerations of technological innovation leads us into the areas of psychology, organization and management science.

6.3 PROMOTING SUCCESSFUL INNOVATION

Important factors relating to success in innovation include the manner of systems design, planning and implementation of change. There are ideas and techniques available which will help ensure success in innovation, by accepting and indeed emphasizing the human and organizational context of the proposed changes. When reviewing technological innovation in manufacturing generally, it must be acknowledged that recent studies in the implementation of advanced technology, especially IT, have raised doubts about the real benefits of innovation. Pay-off failures have been reported, and 'best practice' appears to be rare among users. Many attempts to introduce new technology have not been successful, often due to human or organizational factors. Existing studies suggest that at least 20% of IT investment is wasted, bringing no benefits to manufacturers.

The important factors promoting success in the implementation of change have been examined using post-innovation studies of various organizations where change has been attempted. The results of a study reported by Bessant and Haywood (1986) identify the major organizational factors involved, which are skills, integration, structure, work organization, culture, and information flow. A particular combination of these factors may tend to predispose an organization to accept change effectively.

In addition, it is suggested that many of the benefits of technological innovation have been achieved through the organizational and associated changes required to support the new technology, rather than through the technology itself.

Research results like these have led some people to the conclusion that organizational rather than technical change is the key dimension, with the evolution of new patterns of work organization associated with simplified organization, teamwork, flexibility and devolved responsibility in a less hierarchical structure. Strategies such as TQM and JIT may be seen as expressions of this direction in thought.

One model of a preferred culture has been described as *World Class Manufacturing* (Schonberger, 1986) – the economic manufacture of high product variety with high product quality, with emphasis on closeness to the customer and responsiveness to market trends. It is claimed that the achievement of this goal should be accompanied by exceptional reductions in inventory, lead time, quality problems, space and labour requirements, and lead to real competitive advantage.

For this book, an approach to innovation in manufacturing systems is suggested which is based on one proposed by Hughes and Maull (1985), as a suitable response to problems in IT innovation and the implementation of change. This is a simple planned evolutionary process, linked with ideas in IT strategy, systems design methods, prototyping and user involvement.

At present there is certainly no accepted model of an IT strategy for the automation and integration of quality systems, so, like systems design, it has been allotted a separate chapter which suggests a model approach.

6.4 PLANNED EVOLUTION

Most successful change is planned change. Planning is a fundamental management activity, and planning for innovation is often tackled using the methods of project management, for which special tools and techniques exist, such as critical path analysis. It is not within the scope of this chapter to examine these planning techniques in detail.

Project management methods can be applied to obtain an overview of every aspect of innovation, but, as will be seen below, they are particularly relevant to the implementation stage of the planned evolution process which will now be described.

It is important to have available a practical and straightforward pattern for systems innovation, and planned evolution is a useful structure. Planned evolution requires that considerable effort be spent, during the initial stages of the process, on providing a framework for systems development, ensuring that the resulting system is maintainable and extendible. The approach suggested here makes the assumption (which is not always applicable to greenfield innovation) that proposed systems will evolve from existing systems.

It also assumes that a prerequisite for the design of successful systems is a clear understanding of existing systems and a specification of the proposed system requirements developed in consultation with the system users.

The planned evolution process has four stages:

1. description of existing system, where models of the existing system are constructed;
2. design of proposed system where models of the proposed system are constructed in consultation with users;
3. comparison of the existing and proposed system models to identify the differences;
4. planning and execution of an implementation programme to move from the existing to the proposed system.

The first two stages involve the use of system models. All system modelling is carried out to provide systems descriptions (usually simplified) which are constructed to aid understanding. Models may take a number of forms. Many developments will require several different types of model. The most important types for quality systems innovation are outlined below.

Mathematical models depict the operation of systems mechanisms using mathematical equations and functions which behave in more or less the same way as the real system. A break-even analysis (such as the one used in Chapter 4) is an elementary mathematical model. Another model common in QC is the normal distribution, used to approximate the statistical distribution of a quality characteristic from a manufacturing process. Hard systems and OR techniques are usually based on the use of mathematical models, and many complex models have been devised to describe the behaviour of manufacturing and quality systems.

Simulation models represent the dynamic function of a system during operation, and are often based on mathematical models, such as those of queuing theory. Many types of manufacturing simulation technique are available with computer support. Generally, the software comes in the form of simulation languages or simulation packages. Languages tend to be more flexible, but require a greater degree of expertise to produce results. Packages produce results more readily for the inexperienced user, but are less easily tailored to unusual applications. Both can present a graphical display of system elements and product flow through the system, and can predict statistics such as throughput and machine utilization and their financial implications.

The simulation of quantity issues in manufacturing is the main purpose of most simulation software. Bottlenecks in production flows can be detected, and changes made to investigate the effects which will result. The techniques can also be valuable for quality matters. Inspection stations as well as production processes and storage facilities can be defined. This allows the effects of different diagnosis intervals, inspection times, rejection rates and even the distributions of quality characteristics to be investigated. Simulation can be useful if investment in automated inspection equipment is under consideration.

The greatest advantage of dynamic simulation methods is the ability to make predictions without experimentation on the real system, with all its attendant costs and disruption. Unfortunately, because simulation packages are not directly aimed at the quality area, they lack some of the features which could make them really valuable. In any case, the scope of the models is limited, because they can offer little assistance in many important areas of the innovation process, such as the information, human and organizational implications.

Graphical or diagrammatic models include such traditional techniques as scale models, plans, drawings and flowcharts, and are widely used for all manner of systems descriptions. Many of the computer systems description methods also produce graphical models – for example, data entity diagrams to represent the contents of databases. All produce static or steady-state models, and hence cannot represent the dynamic behaviour of a system.

One useful graphical technique is input-output analysis, which is a way of defining the necessary ingredients needed to give the required results. The activities and processes are drawn as boxes, the required systems outputs (such as material and information flows) listed, and finally the necessary inputs identified. Inputs and outputs are shown as arrows. The diagrams for each process can be linked together to show where outputs from one process are used as inputs by another. Each diagram must be carefully defined, particularly in terms of its interfaces. Input-output analysis software tools are available, and the technique forms part of some systems design methodologies. It is poor, however, at dealing with the complexity of real manufacturing and quality systems.

Another widely used technique, which has already been mentioned, combines an effective method of dealing with complexity with an underlying philosophy attractive in its approach to the relationship between systems and their users. IDEF0 is a structured analysis technique – a graphical tool for creating and maintaining a dynamic dialogue on the type of systems required. IDEF0 has been used to develop and refine models of quality and other manufacturing systems at a number of well-known companies including Du Pont, ICL, IBM, Rolls-Royce and Garret Automotive. In the United States it has been used to describe the manufacturing systems of a number of major suppliers to the US Air Force. Chapter 7 describes the use of IDEF0 for modelling both existing and proposed quality systems.

Once suitable modelling methods have been selected, the planned evolution process begins with the definition of the existing situation. This allows a full appreciation of the current position to be gained as a basis for the next stage, the design of the proposed system. It is often the case that the data collection required for the description of the existing situation uncovers information on problems and other factors which had not previously been fully understood.

The design stage must start with a definition of the objectives of the new system. These may be financial or other quantitative goals, or may be expressed in the form of a mission statement. The design process can then proceed, often using static graphical methods first, followed by simulation to represent the systems dynamics. The resulting model must be a complete description of the proposed system, incorporating all the required detail, such as the procedures needed, control and information systems specifications, and the necessary integration with other systems.

In practice, the design will often incorporate substantial elements of the existing system. This makes it easier to compare existing and proposed systems models and identify the 'planning space' – the differences for which an implementation plan must be prepared. The implementation can then be planned using the well-known techniques of project management for which software tools are available.

The use of a planned evolutionary process may seem tiresome and restrictive at times, but it supplies a structure and discipline to innovation which is of great value. The use of planned evolution must be combined, however, with a coherent systems strategy, with user involvement and an appreciation of the value of a prototyping approach.

6.5 SYSTEMS STRATEGY

A comprehensive systems strategy is essential to successful innovation, particularly in computing and IT. *Ad hoc* developments are only likely to give automation and those who try to introduce it a bad name! The development of a strategy requires an appreciation of information requirements and ideas in systems architecture. Particularly important is the need to develop computer systems capable of incremental evolution as part of a policy of never-ending improvement.

Systems architecture comprises issues such as the method of data representation, the system communications, integration with other systems, the user interface, interfaces to sensors, and systems configuration. This last area is one of the fundamental strategic questions, and refers to the overall layout and shape of the system. The configuration may be considered at three different levels: logical, software and hardware.

At all these levels, configuration may tend to be distributed (spread out) or centralized. Distributed configurations have particular advantages. Logical distribution (where systems functions are distinct and autonomous) is a prerequisite for the incremental evolution of systems, and hence is better suited to a dynamic manufacturing environment and to a policy of change and improvement. The opposite situation, logical centralization, would occur if quality data management were designed to be fully integrated into existing production management software. This would mean that the quality function could not be substantially altered and improved without major changes to the other modules, a potentially difficult and expensive process. Many manufacturing software suppliers still see CIM as a centralized monolithic structure, naturally based around their own product range, which is always in the process of expansion to incorporate all possible CIM activities.

Distributed configurations of software and hardware are also valuable, because as well as easing the implementation of future changes they encourage a sense of user ownership of IT resources, which is important if systems are to be improved by, or with the cooperation of, their users.

6.6 USER INVOLVEMENT

It is widely acknowledged that early and extensive user involvement in the design and implementation of computer-based systems assists the acceptance of new technology. It is important that systems design techniques encourage this involvement, first, so that users share ownership of the solution (and so are less likely to resist its introduction), and second, so that the valuable work experience they have acquired can contribute to the generation of solutions.

User involvement is the only method of maintaining a realistic approach towards the implementation of change, as the practical needs and concerns of the system users constantly impinge on the direction of developments.

6.7 PROTOTYPING

It is widely accepted that a static graphical model or written specification of a computer-based information system, although fundamental to the systems design process, is less than adequate for appreciation of the dynamics of a system. On the other hand, working models (such as dynamic simulations) and prototypes form an effective vehicle for the appreciation and refinement of user requirements.

In the common situation, where users or customers cannot precisely specify their requirements, a design strategy involving the development of prototypes can be valuable. Based on the experience gained in the operation of initial prototypes, improved requirements can be established. This facility is particularly useful in novel system developments where the prototype can provide an opportunity for the emergence of new requirements as users become aware of potential system characteristics. There are various types of prototype system, and the appropriate form of prototype depends on the situation.

The adoption of both experimental and exploratory prototyping forms has been found to give valuable structure to innovation in several studies. Exploratory prototyping, or 'prototyping the specification', is aimed at assisting the establishment of user requirements. It is particularly valuable when attempting to extend automation into relatively new areas. Experimental prototyping, 'prototyping the design', where the feasibility of the proposed systems is tested by evaluation in pilot use, is also important to new applications.

6.8 SUMMARY

This chapter has introduced a practical approach to innovation, suitable for the automation of quality systems. The problems which have been experienced with 'naive' manufacturing technology and IT automation, where human and organizational factors have not been given due consideration, have been discussed.

A planned evolutionary process has been proposed, with four stages, involving the modelling of existing and proposed systems, the comparison of the two descriptions to identify the planning space, and finally a planned implementation phase.

The importance of the adoption of a coherent systems strategy, and the value of prototyping and user involvement to successful IT innovation have been emphasized.

Chapter 7 describes and evaluates IDEF0, a powerful modelling method for use with the planned evolutionary process, which incorporates user involvement within its systems development cycle.

7

Quality systems design and improvement

7.1 INTRODUCTION

One of the main conclusions developed in the last two chapters was the importance of the use of an appropriate technique for the design and improvement of quality systems. The chosen method is described now. To confirm the practical effectiveness of the approach, design studies for quality systems automation have been carried out by the author at Du Pont and Rolls-Royce manufacturing plants. This chapter describes the quality systems design process, and draws conclusions concerning the use of IDEF0 for design and improvement of quality systems. First, however, the usual methods of quality systems design will be reviewed.

7.2 CURRENT METHODS

The objective of much quality systems design or revision is to devise a system to comply with national or international quality systems standards such as the ISO 9000 series, or with major customer requirements. A considerable challenge in the design of quality systems is the requirement to meet these objective external standards with a system which is distributed throughout the many business functions of the organization, and in which quality-related activities will be carried out by management and staff in many different departments. To meet this requirement the quality policy of the organization should be decided at the highest level, then expanded and documented in the form of a quality manual, which is a guide to procedures and an aid to decision-making at all levels. Dealing with the complexity of the system description in a large company is a problem, and the quality manual may become too large to be a useful document if it is not split (or 'decomposed') into manageable portions relating to individual departments or job functions.

Quality systems design and revision are usually carried out in a relatively unstructured manner using a variety of *ad hoc* techniques such as written descriptions of procedures and the listing of required functions, with perhaps a diagrammatic technique such as flowcharting for complex information flows. The problem of complexity is frequently tackled by a more-or-less formal top-down decomposition approach to the production of the quality manual. A common method of top-down decomposition is the delegation of the production of part of the quality manual to the management and staff in the appropriate area. This method has the obvious disadvantage that the total quality systems analysis function may be less than adequately performed without some formal means of coordinating the design effort, and without appropriate system design tools to ensure consistency.

The successful design and improvement of systems requires an appreciation of fundamental systems concepts such as those described earlier, and probably also the adoption of an explicit innovation approach, such as the planned evolutionary process introduced in Chapter 6.

7.3 SYSTEMS DESIGN

Because it is important that the design of automated quality systems is compatible with CIM, the designer should be aware of two widely accepted concepts in systems design.

7.3.1 *Top-down and bottom-up approaches*

A top-down approach means that the designer looks first at the entire system in terms of its objectives, sub-systems, functions and environment, before proceeding to the detailed design of any part.

A bottom-up approach, by contrast, involves the alteration and improvement of systems driven from the most detailed level. This approach is considered vital in a TQM culture, because it is the only way to involve everyone in the improvement process. Because most employees have little conception of wider business systems issues, the adoption of a totally bottom-up approach involves a risk that the changes made will not always work to the benefit of the organization as a whole.

A top-down approach should allow the design and improvement of systems which are orientated towards company rather than departmental goals, thus helping to reduce inappropriate sub-system behaviour and conflict between sub-systems. The existence of an outline top-down systems description should help ensure that a local process of evolutionary change and improvement does not result in problems for the larger business system. Unfortunately, the time required to complete a detailed top-down description of an existing system usually means that the system will have changed before the description is finished. The detailed top-down approach is most useful for the design of future systems when innovation is required.

The best evolutionary approach combines top-down design with a bottom-up improvement process, provided they can be married together in the middle. For this purpose a common systems modelling technique which can be used by all is invaluable.

7.3.2 *Modularity*

Breaking down complex systems into modules that may be implemented independently offers two advantages. First, modular design as part of a coherent development strategy provides an effective means of dealing with complexity, because each individual module is small enough to be understandable. Second, by ensuring the separation of functions and software into discrete units, it facilitates systems development and improvement in the future.

A modular design approach provides a framework in which implementation can proceed at a pace appropriate to the ability of the organization to absorb change. A strategy which minimizes the interaction (or 'coupling') between modules enables the modules and applications to be added or removed as part of the evolutionary development of the system. In a situation where a process of never-ending change and improvement is part of company culture and in which future customer requirements are uncertain, this is essential.

7.4 THE IDEF0 STRUCTURED MODELLING TECHNIQUE

ICAM Definition Method Zero (IDEF0) was developed for the ICAM programme of the US Air Force. ICAM was established with the objective of improving the productivity of aerospace contractors by the systematic application of computer technology. The programme was initially aimed at sheet metal fabrication and has involved funding in excess of $100 million.

IDEF0 is a graphical modelling method used to produce functional descriptions of a system, and is a component of the IDEF methods which also include IDEF1 (used to produce information models, especially useful for software design) and IDEF2 (a dynamic modeller for simulation purposes). IDEF0 is one of the most widely used structured modelling techniques. It was originally known as Structured Analysis and Design Technique (SADT) and was developed by Ross (1977). The IDEF0 modelling technique is increasingly used for the description, design and development of complex manufacturing systems. It has, for example, recently been adopted by IBM as an internal world-wide standard. The modelling process has two purposes: to facilitate understanding; and to provide a means of communicating that understanding. It is particularly useful in the context of the planned evolutionary process described in Chapter 6.

A considerable body of published material describing and evaluating the use of IDEF0 in manufacturing and quality situations is available. Hughes (1983) has examined the use of SADT, and Maull (1986) has provided an analysis of the use and effectiveness of IDEF0 for the description of CIM systems, on which much of this appreciation of IDEF0 is based. Stephens and Fox (1987) have described applications in quality systems design. Several researchers working with industry have used IDEF0 for the modelling of quality systems, notably Maull (1988), Marsh (1988) (who used the technique at ICL and refers to it as 'Quality Improvement Methods Analysis' (QIMA)) and Crossfield *et al.* (1988).

IDEF0 syntax will be readily understood by anyone who has used block diagrams to represent a process or system, as the technique uses boxes coupled together with arrows in a similar way. Processes, functions or activities are represented by boxes on a diagram, while the arrows that connect to a box represent 'data' – objects or information needed or produced by the activity. Four different types of arrow are used in IDEF0:

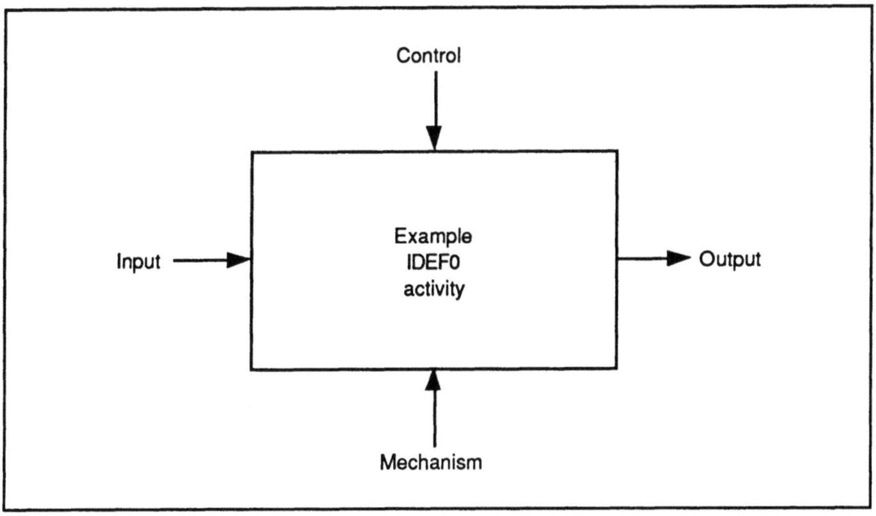

Figure 7.1 IDEF0 activity with associated 'data'

1. Input – enters the box from the left-hand side. This arrow indicates the objects or information used and transformed by the activity.
2. Output – leaves the box from the right-hand side. This arrow represents the objects or information which are the outcomes of the activity.
3. Control – enters the box from the top. This arrow describes the constraints, circumstances or procedures which govern the transformation of the inputs to the outputs.
4. Mechanism – enters the box from the bottom. This arrow represents the agency (perhaps a person or device) which carries out the activity.

Figure 7.1 shows an IDEF0 activity box. Multiple arrows of all types may be associated with activities, and they may branch out or join. All arrows are given a label describing the data involved. Full details of the IDEF0 technique are available in Ross *et al.* (1979) and CAM-I (1984). IDEF0 is not a proprietary product, and can be used by anyone.

Perhaps the most valuable aspect of the IDEF0 technique is the expansion of the contents of any activity box shown on a diagram into a lower-level diagram. This process is known as 'top-down decomposition'. To deal with the complexity of real systems, top-down decomposition can be used to break up the complete system model into individual diagrams, each of which can be easily understood. IDEF0 is thus a top-down design method.

The number of activities on any one diagram is normally limited to between three and six. More information than is displayed on a six-activity diagram cannot be easily grasped by the mind.

A 'parent' activity and the diagram detailing it (the 'child') must represent exactly the same part of the system, and hence the external interfaces in the form of input, output, control and mechanism arrows should match. This ensures that the top-down decomposition process is properly performed.

The design process is meant to proceed by means of a formal 'author-reader cycle', where models are passed between the systems designer and the reader (usually a system user) for detailed comment and revision, until all conflicts are resolved, and a joint viewpoint is established. This iterative process encourages a creative dialogue between systems designers and users. The process effectively involves users in the systems analysis and design, and should promote a sense of ownership of any proposed system. The author-reader cycle seems, subjectively, to lead to improved design solutions; however, suggestions that the process can actually optimize system designs have yet to be demonstrated.

The definition of a 'viewpoint' for the description, and a systems boundary and purpose for each model, are essential preliminaries to the process of preparing diagrams. Experience confirms that if these are not adequately and consistently defined at the outset, problems will arise later in the preparation and decomposition of the model.

The purpose of the IDEF0 diagrams shown here is to illustrate the way in which the modelling of a quality system is carried out, showing decomposition of a top-level diagram into more detailed diagrams. The system boundary encompasses the quality system only, and does not show any activities related to manufacturing or other business sub-systems. The viewpoint is that of the manager of the quality function. Figure 7.2, the highest-level diagram (numbered A-0) shows the proposed QA system as a single activity box, with all its associated external data arrows.

This single box is decomposed in Figure 7.3 to the A0 diagram, showing the four principal activities which are carried out, with the data flows which link them together. These activities are:

1. Plan quality programme;
2. Implement quality programme;
3. Perform quality programme;
4. Evaluate quality programme.

Activity box 3 is then further decomposed in Figure 7.4 into the A3 diagram to show the five activities of which it is composed, namely:

1. Improve quality;
2. Assure suppliers;
3. Appraise quality of manufacture;
4. Control non-conformance;
5. Maintain quality database.

Further decomposition of each of these boxes can be carried out, showing an increasing level of detail, as in Figure 7.5. The final diagram shown here (Figure 7.6) represents the activity of 'Computer-assisted inspection' (A331).

A careful perusal of these diagrams will show the reader that the creation of an IDEF0 model is not a trivial task, and can make intellectual demands on both authors and readers which are not appropriate for every potential user. Crossfield (Crossfield *et al.*, 1988) has simplified and adapted the IDEF0 conventions and syntax, and successfully encouraged many shop-floor employees at Garret Automotive where his study was conducted, to use the simplified version (which he initially called IDEFc, and has now re-christened QMAP) themselves, for quality systems and problem descriptions.

Despite this success, there is a widely held view that the constraints imposed by the technique are its major strength as a rigorous design and communication aid for complex systems. The fundamental constraints are the limited amount of information on each diagram, and the need to ensure that diagrams are decomposed in such a way as to be logically consistent. Without these constraints the diagrams are of limited use in the description of complexity, although they may still form useful explanations to aid understanding.

These valuable constraints on the modelling process must not be confused with such presentational matters as the size and exact layout of the diagrams, and the formal requirement for an accompanying descriptive text page for each diagram. The last is valuable for the documentation of large systems, but represents a considerable overhead in terms of preparation time. The advantage is that the documentation chore is performed as part of the design process itself,

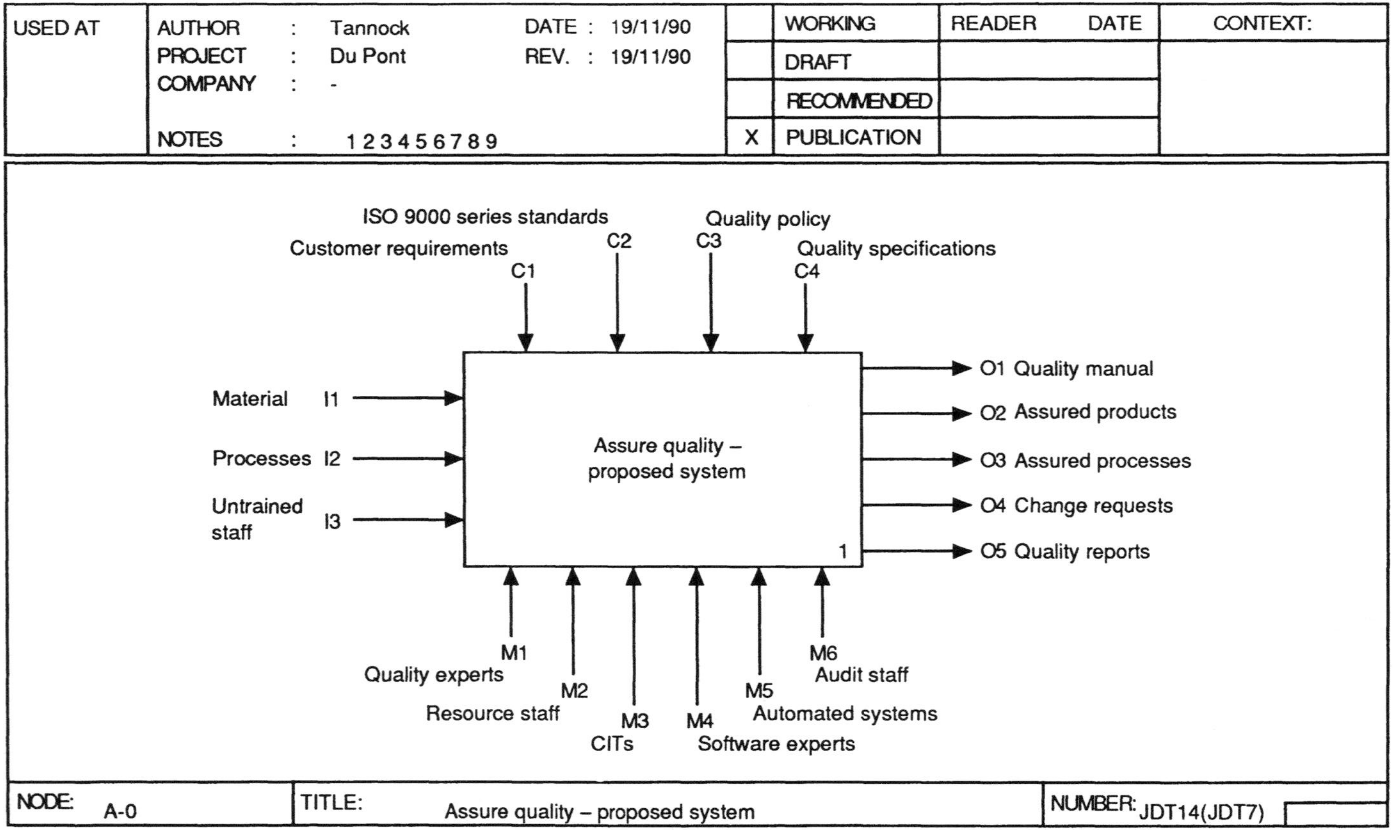

Figure 7.2 Assure quality – proposed system (A-0)

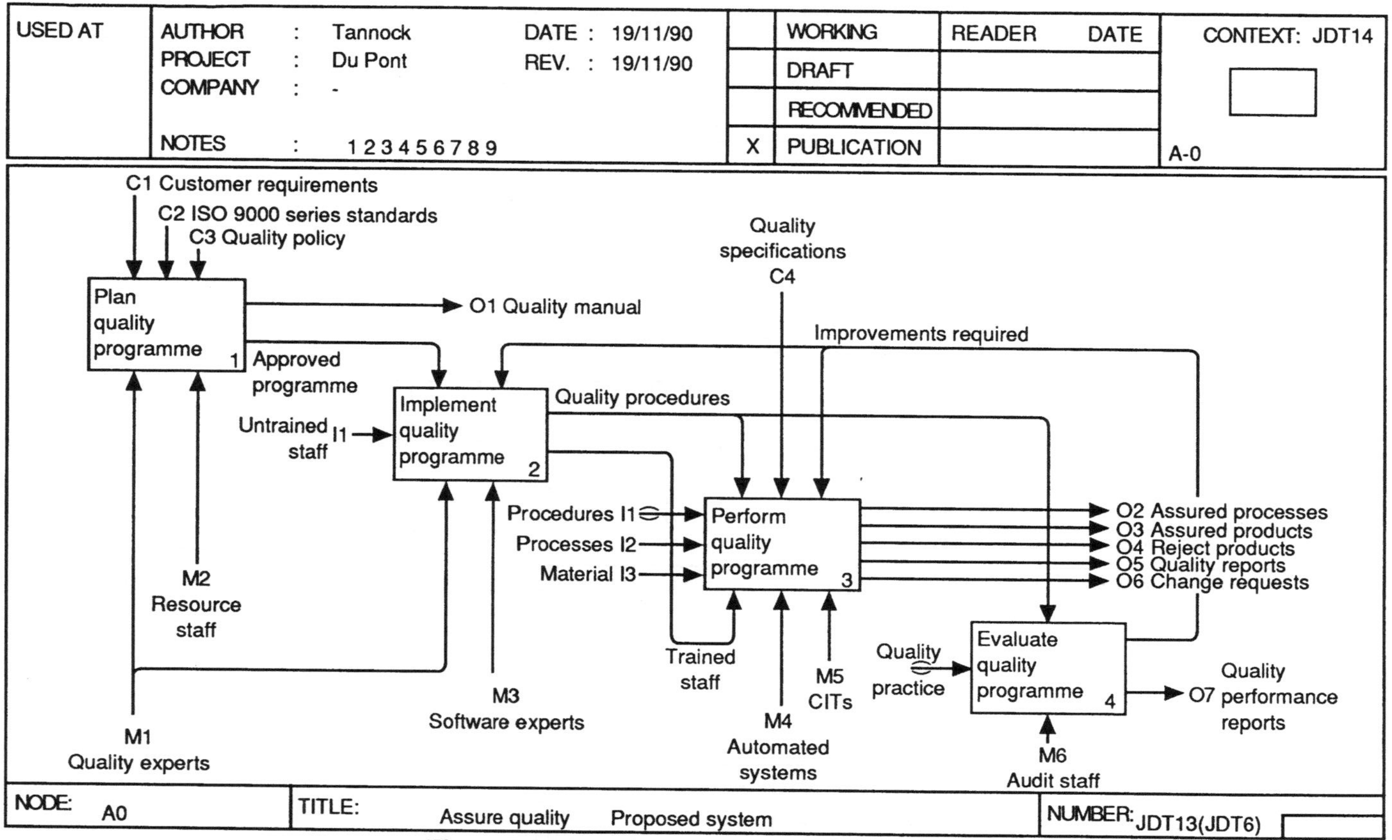

Figure 7.3 Assure quality – proposed system (A0)

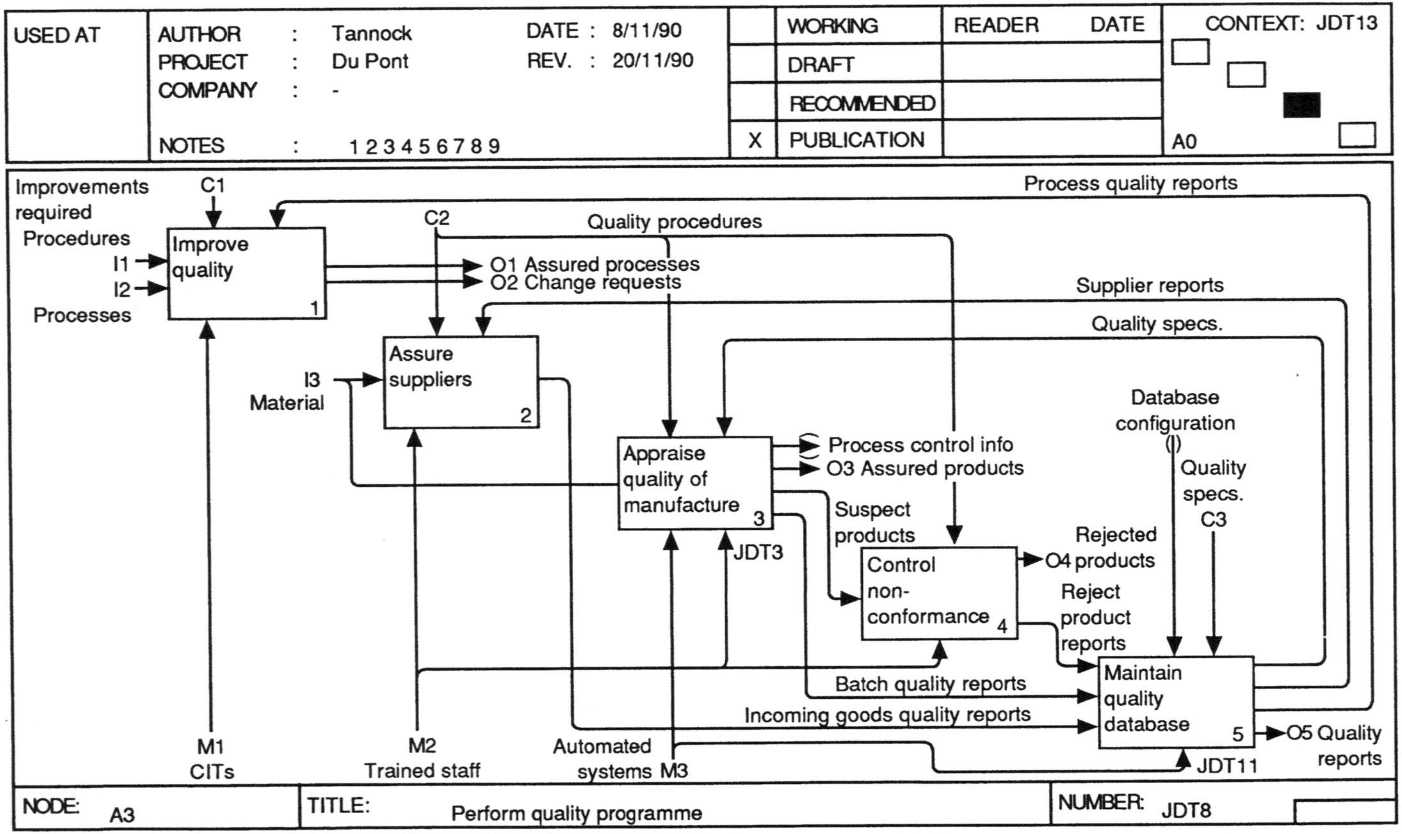

Figure 7.4 Perform quality programme (A3)

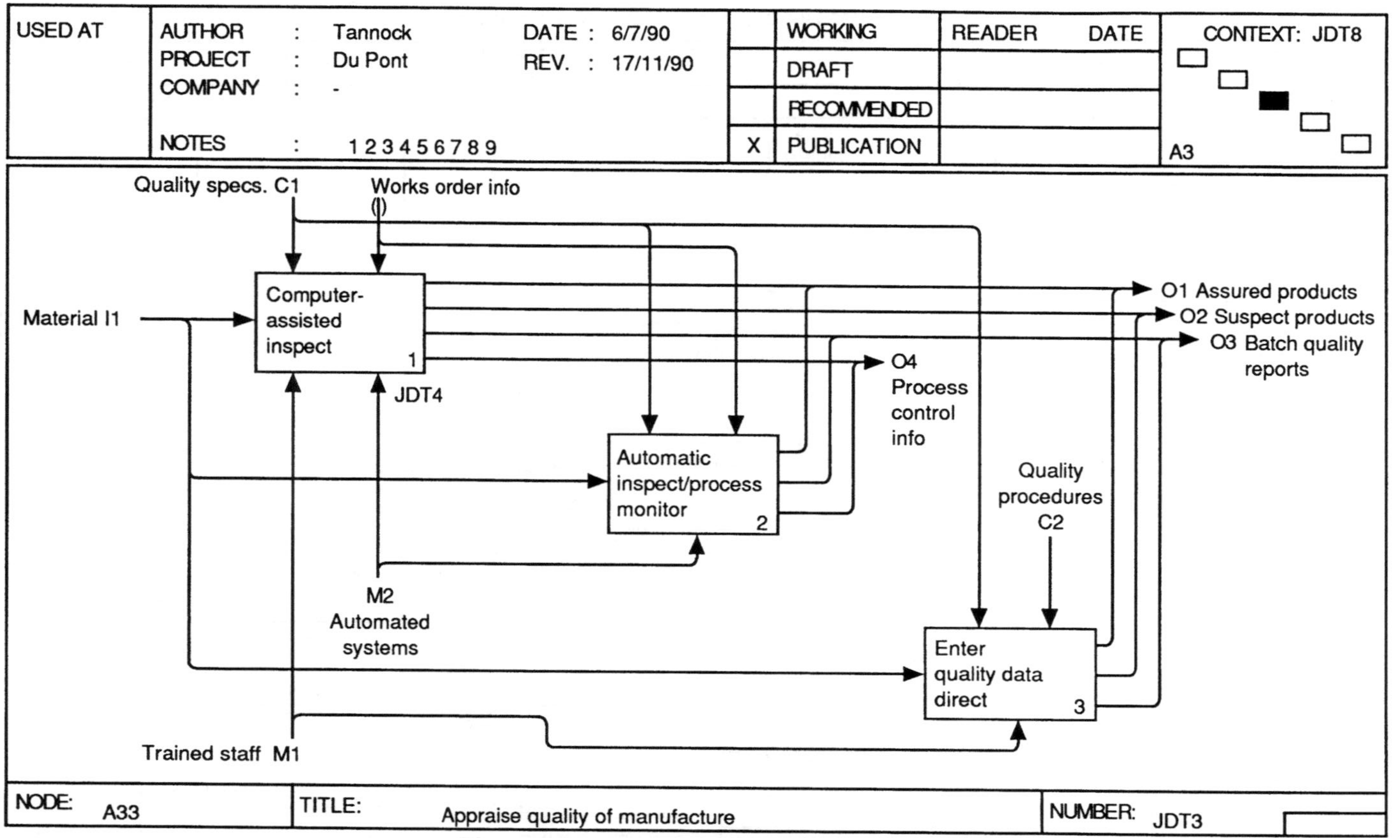

Figure 7.5 Appraise quality of manufacture (A33)

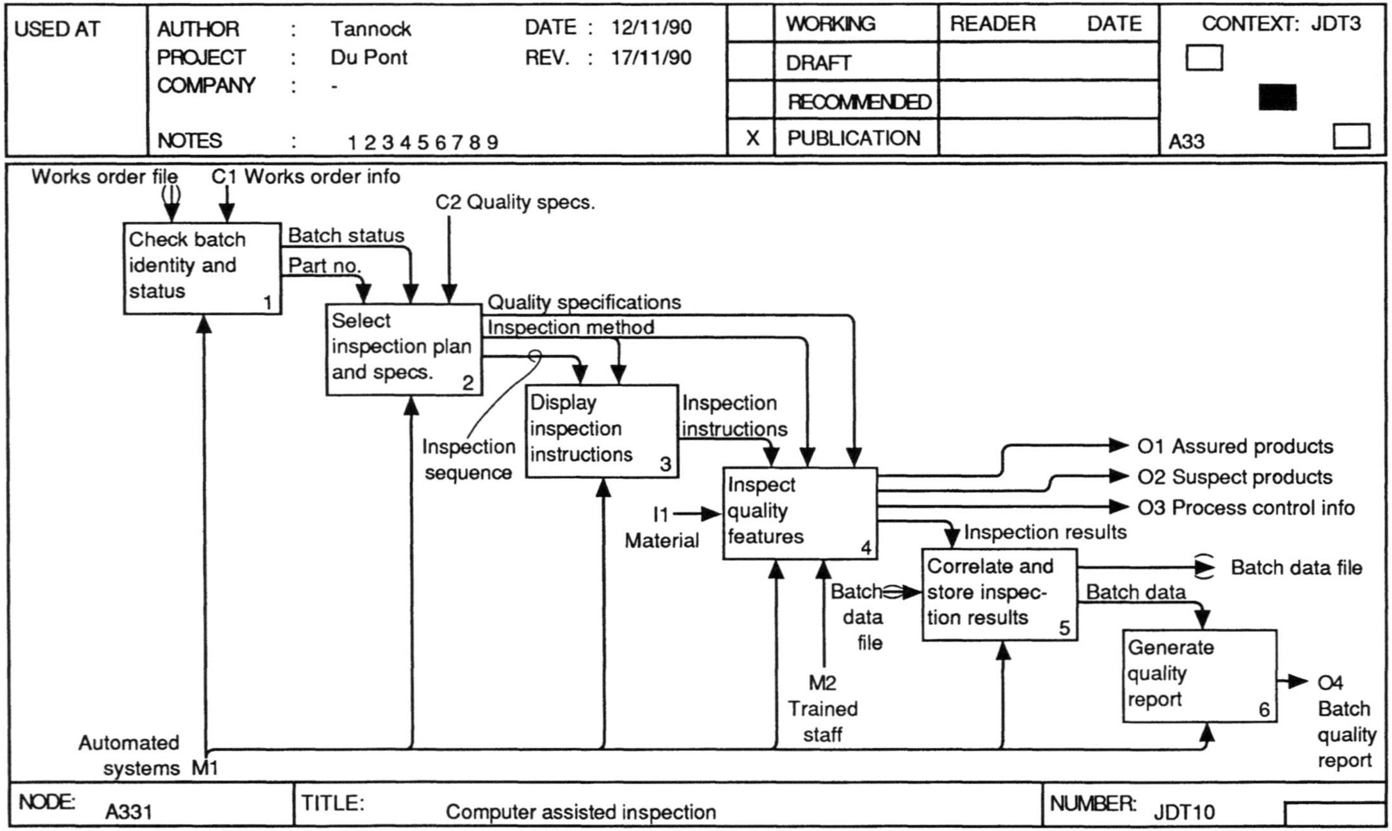

Figure 7.6 Computer-assisted inspection (A331)

which means that a further effort to document the design after it is agreed is unnecessary.

As well as their use for formal top-down design, IDEF0 diagrams are also useful as general aids to understanding systems mechanisms and processes, and for this purpose stand-alone diagrams known as 'for exposition only' (FEO) diagrams are used. FEO diagrams do not form part of a system model, but are used to help explain a particular area by isolating it from associated detail. Anyone who has learnt IDEF0 will find himself or herself using the technique almost instinctively to describe and understand complex situations.

7.5 THE PRACTICAL APPLICATION OF IDEF0

7.5.1 *A software modelling tool*

Perhaps the main difficulty encountered by IDEF0 users is the laborious nature of the modelling process for the complete description of a complex system. The number of diagrams to be developed expands rapidly as the level of detail is increased. A software tool to aid the modelling process is required. A number of products have recently become available for IDEF0 modelling, such as AutoIDEF, IDefine and Design/IDEF. The diagrams in this book have been prepared using IDefine, a professional software system which is available for use with a variety of workstations and PCs. As with all useful IDEF0 software tools, diagrams can be interactively edited and automatically checked for logical consistency. With modelling software as an aid, the potential of IDEF0 as a systems design and improvement tool is greatly enhanced. The author-reader cycle is facilitated by the speed and clarity of the diagram presentation. In addition to the formal author-reader cycle, another technique can be used to develop models for the quality systems design studies. An initial diagram is prepared by the author, based on a preliminary interview with users. Modifications are then made to the diagrams, using the software, in interactive editing sessions with both author and users present, until an agreed description or design is arrived at. In practice this approach was found to be much less time-consuming than the formal author-reader cycle, but retains the invaluable user involvement.

7.5.2 *The model boundaries*

There are two alternative approaches to the modelling of quality systems, both of which are supported in different publications (Ross, 1979; CAM-I, 1984). The two approaches are:

1. to model the complete manufacturing system incorporating the quality system (the 'factory' approach);
2. to model the quality systems independently (the 'generic' approach).

The former has the advantage of making any connections between the sub-system and the rest of the manufacturing system immediately visible, so that quality is addressed within the complete manufacturing system as an embedded sub-system. The implications of any changes are clear and IDEF0 provides a tool to aid integration. It has the disadvantage that, as quality is not addressed independently, it may be difficult to express quality procedures such as those required for quality system standards. Figure 7.7 is an example of an IDEF0 model using the factory approach. The A0 diagram shows the 'Control quality' activity in the context of the entire operations of a manufacturing cell.

The alternative generic approach deals with quality as an independent module or sub-system. Dealing with quality independently recognizes that quality – unlike other sub-systems such as scheduling, process planning, and design – has explicit and recognized standards. Manufacturers need to comply with these standards in order to have market credibility, obtain third-party certification and provide quality plans to prospective customers. If IDEF0 were to be used for this purpose, the modelling would have to be done independently of the rest of the manufacturing system in order that the detail of the quality sub-system would not be lost within the complexity of the entire system.

The generic approach has the disadvantage that the quality system activities are not represented as being integrated with the manufacturing and business system. This position is in conflict with one of the popular themes of TQM, which seeks to ensure that responsibility for quality is devolved into the organization, becoming the task of everyone. It is only when attempting to engage in a practical systems design and improvement process that the difficulty with this approach becomes apparent.

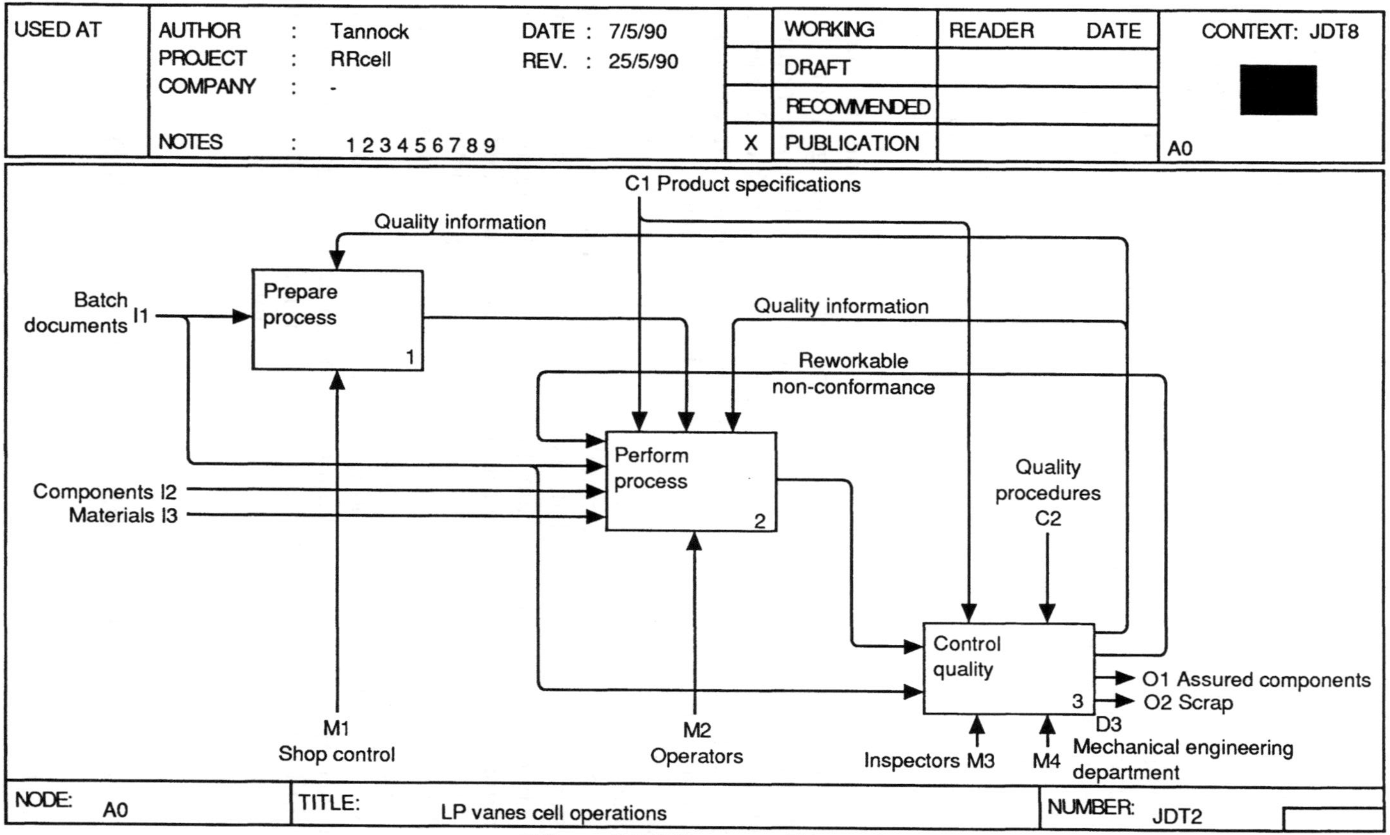

Figure 7.7 LP vanes cell operations

Many organizations hold the view that an effectively defined quality function (expressed in the quality manual as required by a standard for quality systems) must coexist with a TQM philosophy holding that quality activities must be integrated into the organization, so that everyone is involved in quality. This issue again exposes the fundamental duality in quality management thinking between the use of explicit rules and procedures, and the human factors themes designed to obtain the involvement of the workforce in the achievement of improved quality. A systems design and improvement approach must attempt to deal with this potential conflict.

One possible method, while using the generic technique, is to ensure the identification and clear definition of all functional integration linkages between quality and associated business sub-systems such as manufacturing and production control. This can be achieved through the 'tunnelling' of IDEF0 data arrows. Tunnelling is a notation allowing an IDEF0 arrow, shown with brackets around the end, to pass out of view as it crosses a box boundary to emerge later in the outer interface of another diagram, where the data are required. Hence the quality system may be treated separately with each interface being represented by tunnelled arrows to diagrams of other sub-systems. Even if other sub-systems have not been modelled, the exercise forces system analysts and designers to identify necessary linkages at all levels of detail.

Another advantage of tunnelling arrows is found when complex systems are to be modelled, as detailed data flows involved with functional integration can complicate higher-level diagrams until they become difficult to construct and understand. Tunnelled arrows can be used to restrict the information shown on higher-level diagrams, although this practice can conflict with the need for consistency between parent and child diagrams.

While both the factory and generic approaches have drawbacks, on balance it seems that the importance of an accurate description and definition of the quality system as an independent sub-system outweighs the arguments for the modelling of the entire manufacturing system as a whole.

7.6 A CASE STUDY OF QUALITY SYSTEMS DESIGN

A quality systems design exercise was carried out by the author for a Du Pont manufacturing plant. Two suites of models were prepared:

the existing 'as is' quality system and the proposed future quality system. The objective of the exercise was the automation of the existing manual, paper-based QC system.

7.6.1 *The existing system models*

The existing system models showed the quality system from the viewpoint of the system users and administrators. The system users perceived the quality system in a mainly static manner, and although they were clearly aware of the evolutionary changes which were taking place, this was only reflected on the existing system model in the sections of the model which dealt with the introduction of SPC and with the manufacturing quality improvement process.

For the modelling of the existing system the written procedures of the company (devised in accordance with IS0 9001) were a considerable aid. Certain minor inconsistencies in the existing system – areas where procedures had never been fully implemented – were highlighted during the modelling process.

Figure 7.8 shows the top-level diagram representing the existing system.

7.6.2 *The proposed system models*

In the second stage of the planned evolution process, IDEF0 was used to model a proposed IQS for Du Pont. This was to incorporate the ideas of automation and integration in quality data collection, analysis and management which are detailed in other chapters. The author's contribution lay in two areas:

1. developing a coherent and logical framework for the planning, development and evolution of the system, and for the implementation of automation;
2. generating a dialogue with managers and other personnel associated with the system through which future quality systems requirements could emerge.

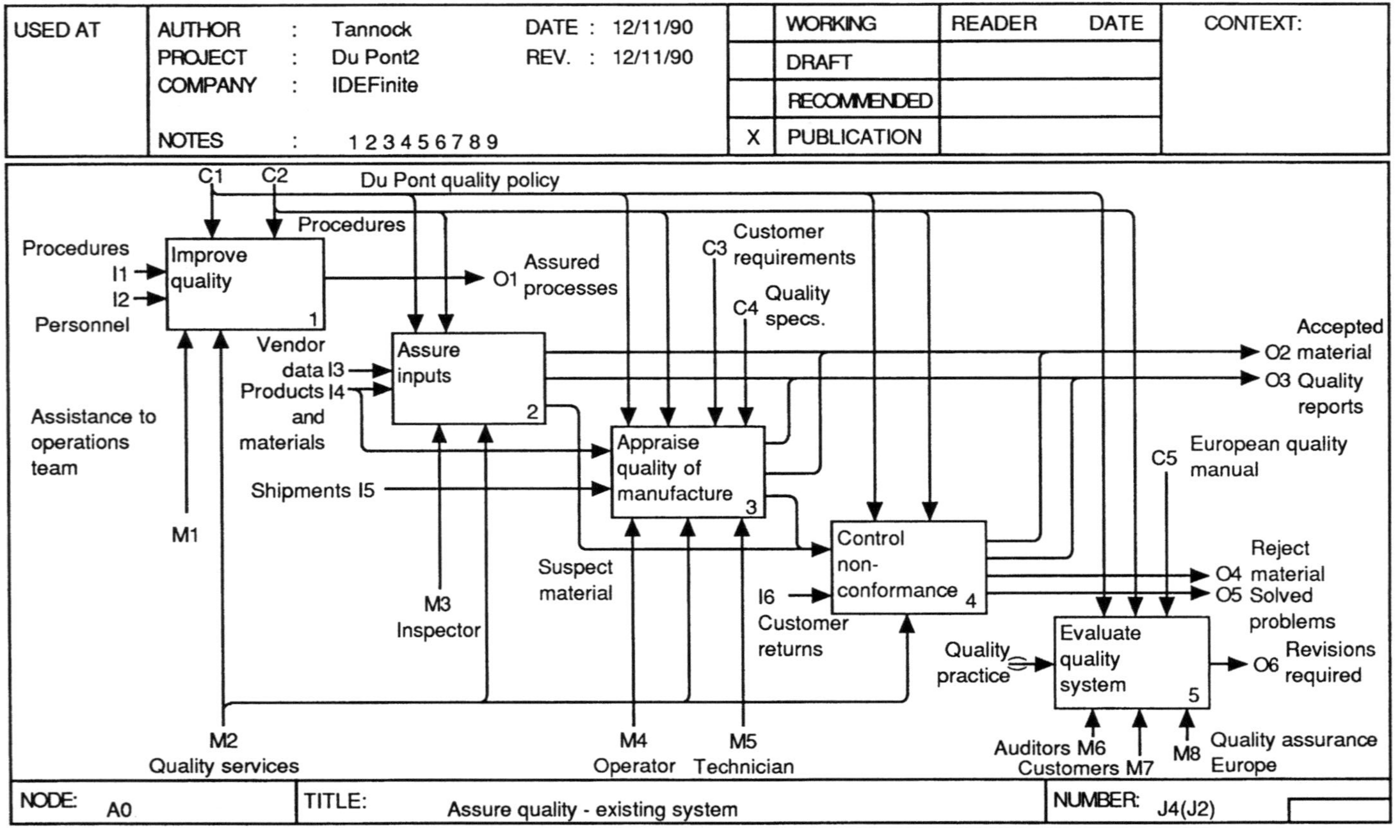

Figure 7.8 Assure quality – existing system

The purpose, boundary and viewpoint from which the proposed IQS models were generated were suggested by the author, rather than by the systems users. However, the existing system models proved very useful in the proposed systems design process, as a basis for developments in thought.

In the areas where automation was practical, the structure of the system reflected the proposed framework for automation. There was no need to consider changes where existing systems (those not amenable to automation) were operating effectively, but in areas where inconsistencies had been exposed by the modelling process improvements were incorporated.

The proposed system model was influenced by CAM-I (1984), which takes a 'quality system lifecycle' viewpoint on the assurance of quality. The planning and implementation of the required systems are brought within the boundaries of the quality systems model. Figure 7.2 shows the diagram 'Assure quality – proposed system' (A0).

It is worth noting that where, as in this case, a planned evolution using IDEF0 was to be the method of design and implementation, the model will describe the process by which it was itself prepared.

Taking this life cycle approach meant that actual performance of the quality systems activities, which had formed the bulk of the existing system model, became a relatively small part of the larger proposed system model. The other major change was the transfer of the quality systems auditing function to the top-level diagram, on the same level as the activity 'Perform quality programme'. A viewpoint which encompasses the system evaluation, planning and implementation process and incorporates a feedback loop for revision and improvement is in line with a policy of never-ending change and improvement of systems.

One essential requirement which emerged from the generation of the proposed system models was that users must develop a sense of ownership of the proposed system if it is to have any chance of implementation. This may be achieved if they are given the opportunity to be trained in the IDEF0 techniques, and if modelling software is made available to them, in order that they may effectively design their own sub-systems, perhaps down to a more detailed level than time would allow for the systems analyst alone. The top-down approach forced the resolution of conflicts between sub-system users, with the system analyst acting as mediator if necessary.

The IDEF0 proposed model described a complete system for the assurance of quality, incorporating:

1. an overall life cycle approach aligned with a process of continuous improvement of systems;
2. automated quality data collection and management activities integrated into the system.

The third stage of the planned evolution involved the comparison of the two models, and the identification of the changes required for implementation. It was found to be easy to identify the major changes required to move to the proposed system. They were:

1. implementation of a CQDB and appropriate computer networking facilities;
2. use of automated inspection and process monitoring for automatic assembly machines;
3. use of CAI for most other quality data acquisition tasks, with direct entry to the database used for minor functions.

7.7 INFORMATION MODELLING

ICAM Definition Method One (IDEF1) – now often encountered as IDEF1 eXtended (IDEF1x) – is a technique for building a data entity model, thus helping to design the necessary database. IDEF1 includes a set of graphical rules and procedures for creating information models, and incorporates a development routine. An IDEF1 data entity model provides a definition of the information needs of the system, such as database tables, files, specifications and procedures. Properly used, it ensures that the information can be shared, is not redundant and that the information system is integrated. IDEF1 also includes a data dictionary.

It is disappointing, in view of the association implied by the similar names and the common source of the two techniques, that there are difficulties with linkage between IDEF1 and IDEF0. Although the techniques are complementary, they are not entirely consistent. A comprehensive survey of the IDEF methods has been made by Godwin *et al.* (1987), who conclude that the significant disadvantages of the IDEF1 technique are:

1. size sensitivity – the data entity model is designed from the 'bottom up' and is not decomposed to deal with complexity;
2. the resource implications, which are considerable when the full

technique is employed.

Godwin *et al.* consider that all entities identified in the IDEF1x model should map to the data arrows in an IDEF0 model of the appropriate level. It is unfortunately the case that not all these arrows may be on the same IDEF0 diagram, as different data entities exist at varying degrees of generality in the information system. However, some IDEF software now available allows the automatic preparation of IDEF1x information models from IDEF0 diagrams.

7.8 CONCLUSIONS - IDEF0 AS A QUALITY SYSTEMS DESIGN TOOL

IDEF0 has great potential as a quality systems design and improvement tool, and may be a valuable aid to companies or consultants aiming to design systems in accordance with standards such as the ISO 9000 series. The particular advantages of using IDEF0 for quality systems modelling are summarized below:

1. The author-reader cycle or interactive sessions used to describe and design systems promote a dialogue tending to improve quality systems. The technique is particularly appropriate to form the basis for quality system improvement activities.
2. IDEF0 meets ISO 9000 series systems' documentation requirements for quality manuals. The diagrams may be used for employee training and reference.
3. The diagrams form a quality systems description document which can be used to give vendors and customers a clear understanding and record of the quality system.
4. A guide to the software specification for automation in quality data acquisition and management may be obtained by a study of the data items and activities of an IDEF0 proposed system model.

The disadvantages of IDEF0 are as follows:

1. As a functional modelling technique, IDEF0 does not represent data entities and their relationships in a way directly appropriate for database design, and there are some difficulties with the integration of IDEF0 and the associated IDEF1x information model.

2. IDEF0 does not show time effects or transfer functions, and cannot provide a dynamic model of a system. There is no prioritizing of activities.
3. The apparent complexity of the technique may discourage some potential users, especially when combined with a software tool.

In conclusion, it is apparent that the use of the IDEF0 technique is complementary to a process of never-ending change and improvement. This should ensure that the technique will be widely used in the future.

8

Strategy for quality systems automation

8.1 INTRODUCTION

Successful innovation in quality systems requires a coherent strategy. Without a clear idea of the objectives of the proposed automation, and how they will be achieved, any project will be handicapped, possibly fatally. This chapter will examine objectives, functions and configuration for company-wide quality systems automation. The approach proposed is called the 'integrated quality system' (IQS), and consists of a plan for automation in quality data collection and management.

8.2 THE OBJECTIVES OF AUTOMATION

A key aspect of the quality function in a manufacturing organization is the data management system, which functions to acquire, store, evaluate and present quality information. A mission statement for automation and integration of these activities might include objectives such as:

1. automatically to acquire, manage and provide quality information in the enterprise such that correct, timely information is available for decision support and reporting purposes at all levels.
2. to form a nexus which will facilitate integration of quality information with the other sub-systems of the business, in a move towards CIM.

8.3 THE INTEGRATED QUALITY SYSTEM

The IQS must collect and manage quality information from manufacturing and other areas. A possible configuration for the quality data collection aspect of the system will be suggested in this chapter, while quality data management will be discussed in Chapter 15.

The advantages of a distributed configuration in respect of systems evolution and improvement were described in Chapter 6. In addition to this important factor, automation in quality data collection often requires local processing power, memory and data storage ability. A distributed system (in terms of software and hardware) will tend to reduce data transfer and data access time, through the storage of the required information close to the user (which may be either a person or a computer). This is particularly important when automatic processes must be monitored and controlled in real time.

The advantages of the alternative centralized configuration arise from three areas: first, from the requirement of decision-makers for generalized plant-wide quality information concerned with many production batches, products or suppliers over extended time periods; second, from the need to exercise demonstrable control over inspection specifications, equipment and records, in accordance with the requirements of quality systems standards; and third, from the facilitation of integration with existing company-wide computerized functions due to the existence of a single node with which other CIM sub-systems can communicate.

A centralized configuration implies a hierarchical system. The hierarchical principle means that information of a more general nature is processed and stored at a higher level in a system, and that of a more detailed nature at a lower level. This is in line with the manner in which decisions are taken on the basis of different types of information within a business system. At technician or operator level, specific and detailed process monitoring and product inspection information is required to enable correct setting of production equipment, and to detect non-conformance or damaging trends in the process. At a higher level in the organization, management requires reports concerning quality levels on a time, product, supplier or batch basis, and quality engineers need analyses of non-conformance to enable improvement activities to be targeted effectively.

Because both distributed and centralized layouts have their advantages for quality systems automation, the best solution is to

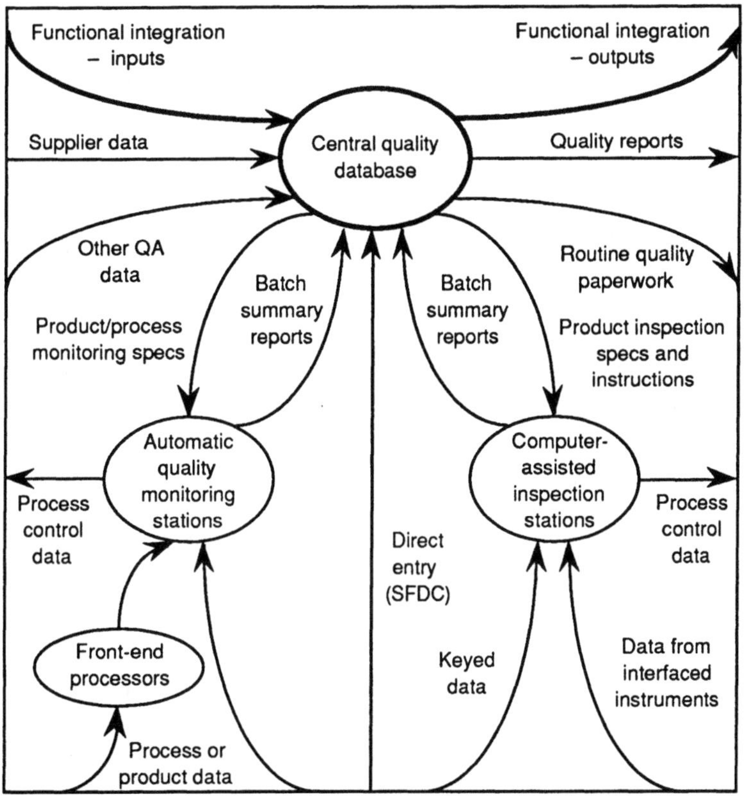

Figure 8.1 Configuration of the IQS

design a hybrid system, with the most valuable features of both hierarchical and distributed layouts. The approach suggested here involves the use of a central quality database (CQDB) to provide the generalized quality data, together with terminals or microcomputer data collection units to deal with the more specific needs for detailed information, mainly from manufacturing.

The logical configuration for an IQS to meet the systems objectives and preferred layout is shown in Figure 8.1. This diagram shows the vertical integration of the quality sub-system, indicated by the arrows representing data communications running up to and down from the CQDB. These are the routes by which quality information is transferred to and from the database. Quality data from manufacturing can be obtained by the CQDB in three ways: by direct entry; by CAI stations; or by fully automatic quality monitoring stations.

The last two methods use a local computer in order automatically to analyse and summarize the quality data before passing them to the database. Detailed process control information is available directly from these local quality computers, which may also have automatic process control functions. Other product quality data, as well as quality information concerning matters such as supplier, customer complaints and calibration of instruments, may be entered directly into the database.

Aspects of functional integration are also shown on the diagram. Outputs from the CQDB to the general business system include quality reports, as well as routine quality paperwork such as certificates of conformance. Most functional integration with other automated CIM sub-systems (such as production scheduling, logistics and design) will involve communications with the CQDB, rather than with the local computers. Integration with the manufacturing sub-system should also be possible at local level, however, where there may be a need for rapid and detailed process control communications.

8.4 QUALITY DATA COLLECTION STRATEGY

Three types of automated quality data collection from manufacturing were suggested above:

1. direct entry, where the inspection task is a manual or visual one, and the quality information is entered via a terminal for transmission directly to the database;
2. computer-aided inspection (CAI), where a local computer gives instructions to a human inspector as to the inspection procedures and use of measurement equipment, and accepts product or process quality information directly from the equipment for local evaluation and recording;
3. automated inspection, where the data collection, analysis, communications and perhaps process control elements of the task are performed without human intervention.

Before discussing these three methods in detail, it is useful to review some of the basic ideas in quality data collection from manufacturing. Quality data can be collected from a number of different stages in the manufacturing cycle: first, from raw materials or components arriving in the factory; second, from the monitoring of processes or the examination of part finished products, either during (in-process) or after (post-process) manufacturing operations; and third, as a final stage prior to dispatch. Automation can be applied to all these stages.

Quality data are collected for a number of purposes: the detection of non-conformance; to allow the control of processes; as an indicator of manufacturing or supplier quality performance; and as an aid to decision support and process improvement. The collection of data can only be justified by their use; there is no point in collecting quality data which will not be analysed, or when no action will be taken as a consequence.

There are two ways to collect quality data from manufacturing: by inspection of the product; or by monitoring of the process. Process monitoring for quality data collection purposes is fairly widely used, but product inspection is almost universally carried out by manufacturing organizations. This is a reflection of the prevalence of tolerance-based design and manufacturing systems. Manufacturing performance-based systems, using operational limits which reflect process capability, have been advocated by some authors, such as Owen (1987). There are substantial barriers to the wide acceptance of such ideas, centred around the functional requirements of component design, especially for assembly.

There are two types of product inspection: gauging and measurement. The word 'gauging' is often used loosely to refer to any quality assessment, but should strictly refer to the use of equipment which detects conformance or non-conformance to product specifications, by indicating if a tolerance limit has been exceeded. Gauging obtains less information than measurement, which can quantify the actual value of a quality characteristic. Effective process control normally requires that measurement techniques are used in order to detect trends before the specification limits are exceeded.

Quality data may consist of attribute data (data referring to discrete conditions, or obtained by gauging of continuous variables) or variable data (referring to continuous variables, obtained by measurement).

There are two basic approaches to product or component inspection: sampling inspection and 100% inspection. Sampling inspection is used where the cost of product appraisal, or some technical difficulty such as process speed versus inspection speed, causes only a selection of products to be examined. There are no other important reasons for the use of sampling techniques, which introduce additional possibilities of error in quality assessment. The only exception is where homogeneous (usually gaseous or liquid) products are concerned. Here representative small samples are all that is required to assess the quality of the entire batch.

Sampling is frequently used for incoming goods, with the help of techniques based on probability theory. For manufacturing inspection, a certain amount of process knowledge is required, often focused by statistical methods like SPC control charting. Statistical techniques for the analysis of both attribute and variable data from samples are well understood, and are usually combined with random sampling to assess the quality of the entire production and evaluate process trends.

Where a process or product is subjected to periodical sampling, an important element in quality data collection strategy is the optimum frequency of the monitoring or inspection, referred to as the 'diagnosis interval'. For sampling inspection, a trade-off of the costs of inspection against those resulting from undetected non-conformance is fundamental to the use of the strategy. Costs are associated with the diagnostic process, the occurrence of non-conformance and the adjustment or recovery activities required. The likely average interval between non-conforming items is important, as is the time lag between detection of a defective item and process intervention to correct the problem.

Taguchi (1981) proposes that the optimal diagnosis interval is to be found by setting the diagnosis cost per product equal to the loss due to producing defective items, and suggests that:

$$n = \frac{2(u + L)B}{A - (C/u)} \tag{8.1}$$

where n is the optimum diagnostic interval, u the average trouble occurrence interval, B the diagnosis cost, C the recovery/adjustment cost, A the loss due to producing a defective item, and L the time lag. Effective automation should reduce the optimum diagnosis interval, by reducing B and perhaps L. The optimum diagnostic interval may then tend towards 100% inspection. Although savings may well result, it is important that the automation of quality data collection should not be seen only as an exercise in cost reduction, perhaps through replacement of human inspectors by machines. The chief purpose of quality systems automation should be to add value to the quality information concerning product and process.

The accuracy and integrity (and hence value) of quality data can be greatly improved by automation in data collection. A description by Kelly (1989) of the improvement of data integrity at an ICL plant shows that without automation in data collection, data accuracy was only 90%. After a seven-year programme of automation, using laser scanners, ATE, barcode readers and SFDC terminals (resulting in about 60% of manufacturing data being collected automatically) the data accuracy had improved to 99%.

100% inspection is currently used mainly where process behaviour is excessively variable compared with tolerances (poor process capability); where process knowledge is inadequate to predict product quality from samples; or where the consequential costs (or other effects) of non-conformance are unacceptable. 100% inspection is also widely used for safety critical products.

An influential theme of TQM advocates the elimination of 100% inspection, and its replacement with statistical methods based on sampling, assuming that the need for inspection cannot be totally eliminated by process improvement. If a manufacturing process has an avoidable problem which is amenable to improvement, investment in the process is preferable to process monitoring or inspection. Sometimes, however, effective process control can only be exercised as a result of process monitoring or product inspection.

SPC methods such as control charts for the detection of trends may also be applied to 100% inspection data. Conventional control charts are set up to cope with the statistical effects of taking samples, by adjusting the control limits using special mathematical factors. All that is necessary to use control charting methods with 100% inspection is to omit these factors.

Both sampling and 100% inspection can be used by all three of the automated quality data collection techniques used by the IQS. Process monitoring and in-process quality control have some additional aspects, and these are examined in Chapter 9.

The need for automation usually arises from a combination of factors involving the manufacturing process and the product. Table 8.1 suggests some of the most important.

The most important factor seems in practice to be a perceived need for 100% inspection. Manual 100% inspection is highly expensive, yet is not a foolproof way of ensuring a high-quality product. Inspectors make errors, suffer from inattention and become tired and distracted, especially when performing routine viewing tasks involving large numbers of identical items.

Table 8.1 Factors in the selection of automatic quality monitoring

1. Process speed	Too fast for manual methods
2. Poor yield rates	100% inspection/sorting needed
3. Complex product	Large amounts of data needed and involved inspection task
4. Variable process	Automatic process control or product adjustment needed
5. Environment	Too hostile for human inspector
6. Failure consequence	Serious consequences of product failure justify automation
7. Quality standards	Exceptional quality requirements necessitate 100% inspection

As regards process speed, the monitoring equipment must be able to obtain sufficient information from the process or product. Quality data are sufficient when they allow immediate process control to be exercised effectively, provide the necessary product quality data for reporting and records, and enable the rapid analysis of quality information for prevention and improvement. The approach which should be taken is influenced by many factors, perhaps the most important being the nature of the requirement for real-time ability in data collection for process control. Real-time data collection may be achieved in a number of ways, depending on the complexity of the data, the rate at which data must be collected, and the other functions required of the system. Communications and data analysis requirements exist in all practical quality systems, and these must coexist with the data collection task.

The decision as to which quality features should be automatically inspected, which level of automation should be used, and whether sampling or 100% inspection will be used, can be made on the basis of a cost-benefit analysis. Higher levels of automation and the collection of larger quantities of quality data cost money, which can only be justified by savings resulting from the use of the information collected. There is no point in collecting information which will not be analysed and used for process control or other purposes.

8.4.1 *Direct entry*

Direct entry involves the manual entry of the quality data to a computer. This is an SFDC operation for quality data, and can be achieved in a number of ways. A terminal linked directly to the database computer is normally used, with VDU and keyboard (sometimes in a simplified form). A number of mobile data capture devices are available, which may be used by a patrol inspector, and these are then connected to the database computer in order to download the quality information to the larger system.

Other methods require the user to fill out a special written form, the data from which are later transferred to the computer by scanning the sheet. One Jaguar Cars application uses a speech recognition system, because the object under inspection is large enough to require the inspector to be mobile. At its most basic, quality data may be entered from a standard inspection report form by a data entry operator, although this introduces an extra possible source of error.

The distinguishing feature of this category is that the user is not positively assisted by the computer in the data collection task, beyond the provision of a data entry screen display or equivalent prompt, and is not provided with assistance in the analysis of the data collected.

Direct-entry inspection elements will form an essential part of a practical IQS. There are many quality data collection and recording tasks, perhaps carried out in response to a particular customer's requirements, which are needed so infrequently as to preclude the development of special hardware and software for more highly automated data collection. The most cost-effective method of obtaining such data for analysis and reporting purposes is through direct entry, via a standard terminal to a database. The design of data entry forms, tables and reports can be quickly and easily achieved using fourth generation language (4GL) database application generators.

8.4.2 *Computer-aided inspection*

The CAI concept incorporates the use of inspection plans and automatic data acquisition from instruments, together with a suitable user interface giving SPC process control information. The computer interacts with the human operator to display inspection sequences and instructions, prompts the appropriate action, inputs and validates data from interfaced inspection equipment and compares it with specification or process capability limits, before providing a report to the quality database.

While the CMM is at present perhaps the instrument most commonly used for CAI, there are many other dimensional measuring instruments available with output which may be used by a computer. CMMs and dimensional gauging equipment are described in Part Three. Other sensor possibilities include temperature, ultrasonic, force, chemical and electrical transducers.

A CAI approach is also used for many inspection tasks in electronics manufacture, especially for functional testing using the wide range of programmable data acquisition systems available for component and product test. Computerized data acquisition systems are widely used in laboratories, and adopt a similar approach. A test program is carried out as follows. The plan is stored in the computer, and when executed, adjusts the product or process to the required state, using the control functions in an interfaced data acquisition instrument. Then the computer directs the instrument to take measurements. These are analysed and a report printed which describes the quality characteristic under test. A sequence of tests can be run under computer control.

Such systems use specialized data collection hardware, and the operating environment and software tend to be designed for rapid reconfiguration by skilled test engineers, rather than for use by process operators.

Another theme in the quality area, which incorporates elements of the CAI concept, is SPC software. SPC software is meant to be used by non-experts, and incorporates a simple user interface, probably allowing data entry from interfaced instruments as well as via the keyboard. Most systems produce batch summary printouts and sophisticated screen displays of control charts to assist process control. Inspection programs and sequences are not usually provided, and the user must choose the critical quality features to study and chart. SPC software is described in Chapter 14.

A situation which should be distinguished from direct entry inspection must be carried out where CAI is being used for the inspection of a component or product with a number of quality features. Many of these characteristics may be inspected using the measurement equipment interfaced directly to the computer, but other features may need special measuring or gauging equipment, which cannot be easily interfaced. Certain tasks will require the operator to make visual inspections and complex quality judgements which cannot be automated. To ensure flexibility, a CAI system needs the ability to accept manual data input of quantitative data, and of qualitative judgements made by the user as part of the inspection cycle.

For example, one engineering firm uses a CMM to inspect machined components produced by a flexible manufacturing system. A major feature of these components is a machined bore, the measurement of which using the CMM was found to be very time-consuming. Inspection of the bore is now carried out much more efficiently using traditional air gauging instruments, and the result entered via the keyboard of the CMM computer.

8.4.3 *Automated inspection*

The availability of cost-effective computing power has made the use of computers for fully automatic quality data collection an obvious step, and a number of technologies have been developed to achieve this full automation of inspection and process monitoring functions.

The purpose of all automated quality data collection is the provision of improved information relating to process and product. Quality data may be immediately analysed and presented to an operator or automatic process controller to allow process adjustment. In addition, information can be made available for component acceptance or rejection. The amount of quality information which may be collected by automatic means is potentially very great, and hence it should be summarized before being passed to a database, which can then generate reports to assist decision-making and process improvement.

Many of the recent advances in automated inspection technology are related to the measurement of machined, cast and moulded components, especially the dimensional inspection of such components within an automated manufacturing environment. The vital requirements are for speed and flexibility in measurement ability, and these have been fulfilled by the development of computer-controlled

CMMs. With the increasing emphasis on small batch production and minimization of work in progress, the flexibility of the inspection equipment must match that of the production facilities.

CMM technology is aimed at the automation of dimensional measurement. Other industries have had to develop their own quality monitoring techniques, with little reference to the metrology traditions which form the foundations of inspection in the metal cutting industries. An example of the sophisticated technology which has been developed for this purpose is found in the inspection and test of electronic components, printed circuit boards and products. Here 'automatic test equipment' (ATE) is the generic name for the equipment used.

Another automated inspection method which has found much favour in a variety of manufacturing environments is machine vision, which may be used for inspection of dimensions and many other quality characteristics. Approximately 90% of the machine vision systems in use are employed to collect quality data.

Returning to dimensional inspection, a well-established category of automated measurement equipment involves the use of dedicated inspection fixtures fitted with dimensional transducers. This type of equipment has a multitude of applications, and is very well suited to integration with an IQS.

Part Three examines these automated inspection and test technologies. In addition, process monitoring and in-process quality control are discussed in the context of their integration into the IQS framework.

8.5 SUMMARY

An IQS has been proposed in this chapter as a strategy for quality systems development, involving the automation of quality data collection and management. System objectives and requirements for systems configuration have been discussed, based on the information, control and integration needs of the organization, and on a recognition of the need for the evolutionary change and improvement of computer systems for quality.

The proposed configuration incorporates a centralized element, the CQDB, together with distributed quality data acquisition and analysis terminals and local computers as required for effective quality data collection. This layout allows a high degree of flexibility for the system, which may be readily reconfigured and expanded. It also provides the security of quality data and control over quality specifications required by quality system standards, and forms a nexus for integration of quality into CIM.

This chapter has also outlined some basic concepts of quality data collection before moving on to consider the three quality data collection methods which have been identified for use in the IQS: direct entry; CAI; and fully automatic inspection. The importance of automation for process monitoring and in-process quality control will be discussed in Part Three.

Part Three

Automatic quality data collection and inspection technology

Automatic quality data collection is a key element of the integrated quality system (IQS). Part Three surveys the technology of automation for both in-process and post-process quality data collection. Should the process or the product be monitored? What is the state of the art, and what techniques are available in a practical and usable form? How may a manufacturer develop successful automation for quality data collection? These questions will be tackled now.

There are several well-established technologies for automatic quality data collection. The best known are described in the following chapters. In addition, there are a multitude of special applications devised to tackle particular products and process. The entire area forms a part of a broad-based move towards automation in the use of computers together with interfaced instruments for the collection and analysis of scientific and engineering data. A wide range of equipment and techniques for scientific instrumentation is available, and those areas most closely associated with automated quality monitoring will be mentioned here.

Automation is common in chemical engineering and the associated process industries, where the importance of automation for primary process control has been recognized for many years. Some manufacturers have progressed through three or four generations of automated process control technology, and now use highly sophisticated equipment to ensure product quality as a result of process parameter control.

Another related area which must be mentioned is condition monitoring, to assist in the maintenance of plant and equipment. In recent years automation in this area has involved the use of sensors and computers to collect and analyse condition data from manufacturing plant. The condition of such equipment will have a more or less direct effect on product quality, and hence this area tends to overlap with automated process monitoring. The chief distinction is the type of information collected, which is concerned with the technical aspects of the plant related to its maintenance needs, rather than with product quality.

Laboratory testing is another field where automation in data collection is increasingly used. Quality control of raw materials and products is the primary function of many industrial laboratories, and automated testing forms one aspect of a trend towards laboratory information management systems (LIMS). In a manufacturing environment, LIMS can act as either a partner or an element of an IQS. The quality data management requirements in LIMS are very similar to those of manufacturing, and at least one commercially available manufacturing quality database system has evolved from LIMS.

Having noted the associated fields, the following chapters will now concentrate on automation in quality monitoring for discrete part manufacture, to carry out those functions traditionally performed by quality control and test staff, and by process operators. The areas which will be examined are in-process quality control, and the post-process inspection technologies of coordinate measuring machines, machine vision, automatic test equipment and dimensional gauging. Equipment from all of these areas can be incorporated into the IQS configuration described previously, as either computer-aided or fully automated quality data collection stations.

9

Automatic in-process quality control

9.1 INTRODUCTION

Process control is carried out to ensure that the resulting product is manufactured to the correct specification. Process control activities, together with in-process product inspection, should both be seen as aspects of in-process QC and therefore part of the overall QC system. This chapter will outline approaches to automatic in-process QC and examine the existing and potential linkages between process control and QC.

All process control requires that either the manufacturing process or the product must be monitored. Process monitoring techniques are as various as the manufacturing processes themselves, and it would be impossible in a single chapter to attempt to cover the entire field. In addition, there are many techniques for in-process product inspection. Hence three representative examples of automation for in-process QC will be described.

Two of these, connected with computer numerical control (CNC) machining and plastic injection moulding, have well-established applications. The machining example described is in-cycle gauging, an in-process product inspection activity. The moulding example represents automated process control, with monitoring of process parameters and output of quality data. Assembly is perhaps a more challenging area, where few existing applications are to be found, and a detailed case study of a process monitoring system development which was carried out by the author is described. This case study illustrates an approach which can be adopted to process monitoring automation within the context of an IQS.

This chapter concentrates on the technical aspects of in-process QC. There are also, of course, motivational, organizational and training aspects to any successful application. Automatic process control technology has usually been seen as the concern of the engineers, managers and operators in manufacturing, rather than the quality function. This has meant that process control and QC have often developed along separate paths, with little functional data integration.

There is, however, a potential advantage of this ownership of process control by manufacturing. It reveals an excellent opportunity to trigger the culture change necessary to make product quality fully the responsibility of those who make the product. The integration of in-process QC with automatic process control can be used to ensure that those in the manufacturing function are in a position to collect, analyse, report and act on their own quality information.

9.2 CONTROLLING THE MANUFACTURING PROCESS

Process control involves the monitoring of one or more process parameters, with some form of information processing to analyse process status and determine any adjustments required. In-process inspection examines the product itself, while it is in an unfinished state.

Both these activities will be followed by a controlling feedback which regulates the process to ensure that the product will be within specification. In this broadest sense, most manufacturing processes are subject to some form of process control. The process controller may be a person, a specialized mechanical or electronic device, or (increasingly) a digital computer.

In the case of a manually controlled process, it is quite common for an operator to attain, through training and experience, a level of skilled performance which results in an excellent quality performance. This is the way of the craftsman, whose personal satisfaction is only complete when the job is perfect. The craftsman's skilled performance makes use of his experience and intelligence to interpret process parameters and product quality information, and make decisions on process adjustment.

The craftsman uses a wide range of product and process parameters to guide his control of a manual process. A skilled turner, for example, uses his micrometer and other measuring instruments to measure dimensions between cuts. He aims at, and usually achieves, a final dimension close to the nominal value. In addition to the dimensional feedback, however, he is also obtaining visual and auditory data which inform him about matters such as material hardness, tool condition, cut parameters, coolant flow, etc. Tactile signals from his hands reinforce the audible and visual information, allowing him to ensure that surface finish is optimized. He can compensate for machine wear, maintain his own cutting tools, and take full responsibility for his own quality performance. He makes use of both in-process inspection and process monitoring data to assess and regulate process performance.

From the production manager's viewpoint, the practical disadvantages of our craftsman turner's process control methods are considerable. He will probably insist on being unhurried in his excellent quality performance. He may object to being expected to carry out repetitive batch work. He will complain at being expected to 'mind' an automated CNC lathe, but may also be reluctant to update his skills to cope with programming the new technology.

By contrast with the craftsman, the majority of manual processes are under the control of semi-skilled and unskilled people from whom very variable process control performance is obtained, leading to product quality problems. Studies of the causes of quality loss indicate that operator-caused non-conformance is a substantial ingredient of the total problem. It is not the only one, however. Many management-controlled factors such as raw material problems, production pressures and inadequate equipment also take their toll on product quality.

It is to cope with the results of all these problems that post-process product inspection has become an almost universal activity in manufacturing.

9.3 IN-PROCESS QUALITY CONTROL OR PRODUCT INSPECTION?

Post-process inspection is only one approach to tackling manufacturing quality problems, and one with major and well-known disadvantages. Inspection is not the way to quality, but a means to manage its absence.

Any effective strategy for product quality must involve the use of an appropriate mix of methods, aimed at eliminating the possibility of unsatisfactory items reaching the customer with minimum additional cost. For manufacturing, the main methods (in descending order of preference) are the following:

1. Process improvement – to eliminate the possibility of defective items. Improvement is aimed at eliminating all important causes of process variability, to ensure that the product is always manufactured within specifications. Techniques such as poka-yoke ('mistake-proofing') have proven worth, especially in areas such as manual assembly. Total confidence in correct process performance will eliminate the need for further control or inspection.
2. In-process quality control. Not every process is amenable to improvement which eliminates all substantial causes of variability. To achieve excellent quality performance in this situation we may employ skilled and highly motivated staff to undertake manual process control or ensure effective automation of the process control function. By monitoring process parameters or by in-process inspection of the product, the information needed to make the necessary adjustments to the process can be collected. The need for post-process product inspection can again be eliminated if full confidence in process control methods is achieved.
3. Post-process quality control. The final resort is post-process inspection or test, to detect and filter out defective items. In many situations there is little practical alternative to inspection. As products and manufacturing procedures become more complex, the possibilities of error proliferate alarmingly. In the foreseeable future, product inspection will be widely used. Inspection does not, of course, have to be 100%. Sampling inspection, especially when combined with SPC methods, is sufficient to control the majority of processes, and to provide vital information for process improvement.

The main advantage claimed for in-process QC over post-process inspection is that the former can (at least in theory) prevent the production of any defective products whatsoever. In-process control should, therefore, be the first choice. However, a well-designed post-process product inspection activity can perform virtually the same control job, but might occasionally allow the production of a small number of identified non-conforming items before adjusting or halting the process.

If in-process QC is appropriate, the next choice is between process monitoring or in-process inspection. The crucial factor with any application of process monitoring is the requirement for precisely known correlation between the process parameter(s) to be monitored and the resulting product variability. In the simplest situation there will be one or more potential process fault(s), all of which are sure to result in definite and specific non-conformance. The first task is to demonstrate that the product faults are associated with definite and repeatable variations in a process parameter which can be readily monitored. Experiments may be necessary to prove that monitoring of the process will allow the product to be acceptable without further inspection.

In a practical manufacturing situation, we must often deal with a sequence of processes, some manually operated, some automatically controlled, with the overall shop control problem compounded by production deadlines, raw material quality problems and a host of incidental breakdowns and irritations. The automated process which can look after itself, control the quality of its output, and bring itself to an instant halt if problems do occur is a boon to the harassed manager.

9.4 EXAMPLES OF AUTOMATIC IN-PROCESS QUALITY CONTROL

9.4.1 *In-cycle gauging with CNC machining*

Perhaps the CNC machine tool is the classic example of automated manufacturing. Like the author, many mechanical and production engineers have an affinity for metal cutting technology (also known as 'proper engineering'), and like to illustrate their reasoning with examples from this area.

Most manual metal cutting machine tools are fitted with handwheels connected to leadscrews which move the cutting tools and workpiece in relation to each other. Powered drives are provided only for the simplest of movements, and these must be started and stopped manually. Our skilled turner uses the handwheels and their associated scales to position the tools for a cut, then often sets the drive to remove the material.

Machine tools have been automated in a number of ways. Early methods used mechanical cams to move the cutting tools into the material. More modern developments use electric motor power, controlled by a unit which reads a 'part program' from a punched paper tape, or more recently (with CNC machine tools) from computer memory. Responding to the codes in the part program, the machine controller sends signals to the axis drive motors, which move the cutting tools or workpiece to achieve the required material removal.

Most CNC machines incorporate 'closed-loop' positional feedback. They are fitted with axis measuring scales – sensors to determine the position of the machine slides. The information from these is used by the machine controller, which compares it with the positions required by the program. If discrepancies exist, the controller will regulate its motor drive control signals to eliminate them.

At first sight, this might appear to be an instance of automatic process control, which would ensure that the correct dimensions were produced each time the part program was run. Unfortunately, much more is needed in practice to ensure specifications are achieved. The exact position, size and shape of the cutting tool are the crucial factors. These must be accurately known to the machine controller, and will be entered into a 'tool offset' table in the controller memory. If this is incorrectly done, if a tool wears or breaks, or if cutting forces cause significant deflection in the system, dimensional errors will result.

Various alternatives exist to ensure that defective parts are not produced. Additional process parameters can be monitored. Noise and vibration from the cutting process can be assessed to determine tool condition or the occurrence of breakage. Spindle motor current is another useful measure of the cutting conditions. But perhaps the most useful monitoring technique allows the tool and work to be dimensionaly examined. This method is known as 'in-cycle gauging', and is an example of in-process quality control.

Touch trigger probes are passive surface-sensing probes, essentially accurate switches which provide a signal when they encounter a surface, and are fitted to machine tools in order to inspect both the workpiece and the cutting tools themselves. (They are also used with CMMs, and are described in more detail in Chapter 10.) The CNC machine tool controller must have appropriate hardware to interface with the probe unit, and the software to take advantage of the information provided. These facilities are common on modern machine tools, albeit usually as expensive optional extras.

In-cycle gauging makes use of the computer-controlled axis moves of the machine tool, normally used to position cutting tools and take cuts, to move a probe which is fitted in place of a cutting tool. This is used in conjunction with the positional measurement facilities built into the machine tool, to assess the position of the surface of a component. The measurement takes place between cuts, and does, of course, increase machining cycle times.

Most modern machine tools have automatic tool changers, or multi-position tool turrets. A probe can, therefore, be called into use in a few seconds, and moved up to the workpiece until the stylus contacts the surface. At this point the probe is triggered and axis measuring scales are read to determine the position of the relevant feature. All this is carried out automatically under the control of the machine tool controller, running a part program incorporating specially written 'macros' (programmed procedures) to obtain and use the probing information.

Information is required before each critical cut to allow process control. Control is exercised by the program macros, which can use the information from the probes to change the offset tables in the machine tool's memory. This technique can compensate for cutting tool wear or component misalignment. Many CNC controllers also allow special-purpose macros and procedures to be written, which can be used to instruct the machine tool to change to an unworn 'sister' tool, output component quality information, alter program execution, or change parameters in the part program itself.

It should be noted that in-process inspection using probes with the machine tool is potentially unreliable, because the same axis measuring scales are used for both cutting and measuring. Defects caused by a fault in the positioning system of the machine tool will not be detected. Independent inspection, using a CMM or automated gauging equipment, is superior in this respect.

Probing is also used for pre-process inspection. Pre-process inspection involves measurements taken before the machining cycle commences, and can be used to assess the precise position and orientation of a workpiece, or the initial size of a casting or forging, to determine the amount of material which must be removed.

Touch trigger probes are also used to sense the position and size of the cutting tools. This will ensure that a worn or broken tool is not used to cut the workpiece. Probes for this purpose are mounted on the machine tool in a position where the tools can be moved into contact with the stylus. The size and condition of the tool are assessed from the position at which it makes contact with the probe, see Figure 9.1.

Some CNC machine tools are now so reliable that with careful attention to cutting tool (that is, process) condition, machined part inspection can be eliminated. One example of such a situation can be found at the fully automated Mazak machine tool factory near Worcester, UK. At this plant, sophisticated cutting tool management

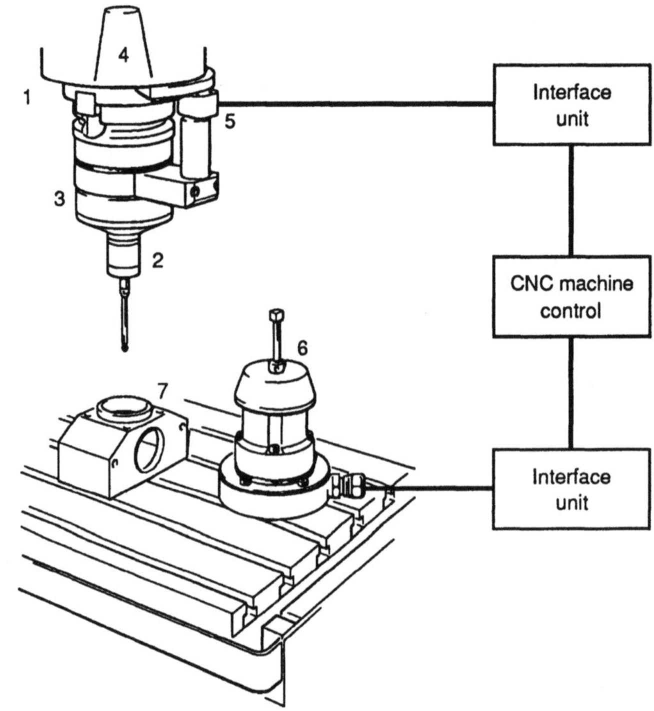

Typical Installations

1. Machine spindle
2. Inspection probe
3. Adaptor
4. Shank
5. Inductive transmission modules
6. Tool setting probe – hard wired
7. Workpiece

Figure 9.1 In-cycle gauging of workpiece and tool
(reproduced with permission of Renishaw International Ltd.)

systems have ensured that the CMMs and other specialized inspection equipment which were designed as part of the flexible manufacturing facility now lie idle.

9.4.2 *Plastic injection moulding*

Injection moulding is straightforward in principle. Thermoplastic granules are heated, mixed and sheared in a screw preplasticizer until they melt, when they are injected under pressure into a cooled metal mould. Several parts may be made at once, by having a number of impressions machined into the mould. The mould is kept closed for a setting period under continued follow-up pressure, then opened to release the component. Moulding machines vary in size from an injection capacity of 12 000 mm^3 up to 2.5×10^6 mm^3. Plastic injection moulded products and components are found everywhere, because the process is fast, convenient and has always been considered economical for medium to large quantities.

Things are changing, however, as they are in other areas of manufacturing. Wortberg (1989), after a detailed survey of many leading plastics manufacturers in West Germany, describes some of the factors involved. There is an increasing emphasis on short runs in a JIT environment, meaning that mould tools must be changed more frequently. A tendency for some large customers (led by the Ford Motor Company with its Q-101 specification) to insist on post-process sampling inspection and the use of SPC methods is also notable. Such customers may require the moulder to take full responsibility for delivered quality direct to assembly lines without a goods inward inspection.

These factors have led moulders to make expensive investment in training and test equipment. There are potential problems with shorter runs in obtaining sufficient set-up data for SPC charting. It is also the experience of some moulders that the SPC techniques appropriate to the metal cutting industry are not necessarily so useful in moulding, as the process variability arises from quite different causes, and there may be difficulty in appreciating the complex linkage between process variability and the quality of the finished article.

Process variability arises firstly from raw material variability in matters such as density and water content. In the moulding machine itself the condition of the melt may be affected by transient conditions on start-up and during breaks in operation, by heater malfunction, ambient conditions and wear in equipment. Maintenance of the correct mould temperature is critical to proper cooling of the parts, and also affects machine cycle time. Different designs of mould have different filling and cooling characteristics.

The injection moulding process is easy to automate, and open-loop controllers (fitted with sequence timers but no process sensing) have been used for many years. Many modern machines are fitted with closed-loop computer controllers which collect data from temperature and pressure sensors, and regulate the process on the basis of this information. So far, we have a typical example of computer process control, with no linkage to QC.

In recent years moulding machine manufacturers have taken a further step, and developed software for their process control systems which can evaluate process parameters and display 'quality screens'. Typical process monitoring variables are:

1. melt temperature;
2. injection, metering and cycle times;
3. melt cushion;
4. pressure integral and maximum value.

The profiles of the process variables may be monitored over each injection cycle, and the actual values compared with the preset parameters. Deviations from the set values can be logged, trigger an alarm, or shut down the process according to the extent by which the process tolerance has been exceeded. Parts with a dubious injection cycle can be identified and segregated automatically. Each cycle can be recorded, and charts produced on-screen showing process parameter variation against time. Statistics of each parameter are also available, and some software can produce SPC control charts based on the process data, with automatically calculated limits which trigger action when a trend is detected.

An examination and understanding of the process parameters can help prove the processing capability of a machine with a particular mould tool, aid fault-finding and improvement of mould tooling, and has been demonstrated to lead to a reduction in mould maintenance costs.

Many controllers now have communication facilities. The author once carried out a feasibility study to examine the linkage of six moulding machines with a microcomputer. Serial (RS-232C) communications were readily established between the controllers and an Apple microcomputer. Process data which had been logged by the controllers could be sent to the micro, which was programmed to archive the data and print process quality reports for each batch produced. A number of moulding companies have developed systems of this type for computerized process monitoring, using enhanced process controllers with quality data linkage to a computer.

It would also have been possible to store process control settings for each product on the Apple, but the machine controllers involved were not capable of receiving their settings from an outside source. Some moulding machine control systems now have this ability. Storing process setting and parameters in a central computer and downloading them as necessary should reduce product changeover times, avoid setting mistakes and improve the ability of the operators to get the product right first time. Different settings will be required for the same mould tool on different moulding machines. All can be stored and used as necessary.

The critical question is whether the knowledge of the process variability provided by enhanced process controllers constitutes an adequate predictor, in practice, of moulded part quality. To what extent can post-process inspection be eliminated? Will customers be prepared to accept statistical evidence of batch quality based on process variables (which they may not fully understand) instead of critical product features? The batch may be demonstrably under statistical control, but how (for example) does a process variable such as pressure integral relate to a vital dimension?

All these questions are as yet unanswered for most moulders. The hardware and software enhancements for moulding machine controllers, to allow them to collect quality data, represent a substantial investment. Only when a moulder and his customers develop full confidence in the process monitoring and control methods available can inspection levels be reduced.

A substantial technical difficulty which holds up progress towards confidence in process monitoring is the sensing of mould tool temperature. This is a critical process variable, especially with multi-impression moulds, but it is difficult to monitor effectively because a temperature sensor may be required for each impression cavity (and perhaps runners as well). This increases mould cost, and requires a customized interface and software configuration in order to connect to the controller. Moulding machine manufacturers do not find this an attractive prospect.

9.5 A CASE STUDY OF ASSEMBLY PROCESS MONITORING

The development of a process monitoring system is described below, as an example of a low-cost special-purpose development which should be within the capabilities of many companies. The system was specifically developed by the author to operate as part of a plant-wide IQS, as a fully automatic quality data collection station.

The reader will need some knowledge of electronics and

microprocessor-based computer systems (and the associated jargon) in order fully to appreciate the following description.

The system was developed for monitoring the high-speed automatic assembly of electrical connectors. This is a very simple assembly operation. The products consisted of moulded plastic bodies with holes into which the gold-plated connector pins were assembled. The assembly machine was a flexible computer-controlled manufacturing cell capable of rapid product changeover. Pins were individually inserted (or 'stitched') at a rate of ten per second from a bandolier, while the plastic bodies were located on carriers which moved along a conveyer system.

A process monitoring approach was chosen because of the advantages of immediate process control feedback, and for reasons of economy. A machine vision system capable of post-process inspection of the product at the required speed had been investigated in depth, and judged practical but uneconomical. Various process monitoring possibilities were then considered, and the assembly force was chosen as the only accessible process parameter. This could be monitored by placing a strain gauge transducer on the mechanism which inserted the pins.

Using a computerized test rig for the investigation of transient conditions as the pins were stitched, it was quickly established that a repeatable 'force signature' existed for the insertion of each type and length of pin, and that the common process faults could be related to changes in this signature. Typical insertion force signatures are shown in Figure 9.2.

Typical quality problems were bent or smashed pins, incorrect pin end position, incorrect pin protrusion or insufficient pin retention force. Defects occurred in two modes: an infrequent and apparently random occurrence when there was no identifiable process or material fault; and an increasing incidence of problems when wear or maladjustment in the insertion mechanism was starting to affect the process.

9.5.1 *System hardware*

The computer hardware used for the prototype system consisted of a microcomputer operating as an automatic quality data collection station, with a specially developed front-end processor which was used

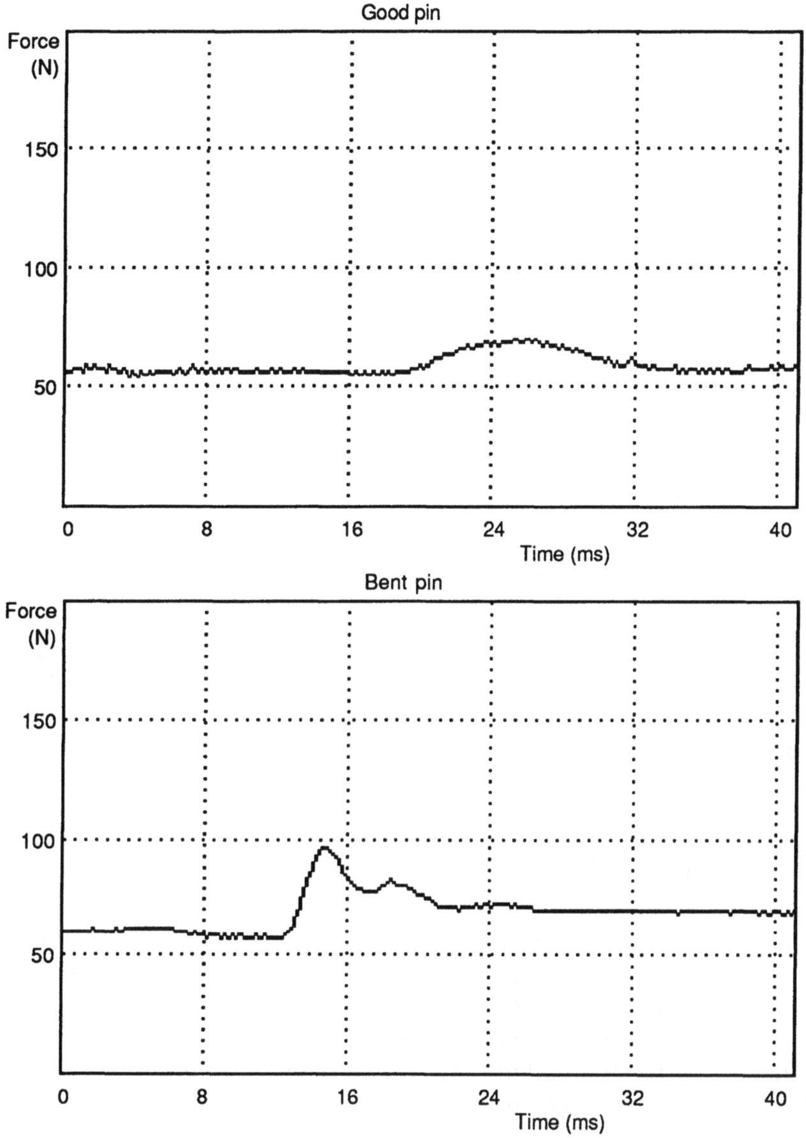

Figure 9.2 Typical insertion signatures

to monitor the insertion force. The microcomputer was an IBM PC/AT compatible fitted with an IEEE-488 interface, the high-speed parallel communications standard widely used for communication with instrumentation. This interface was chosen for communications with the front-end processor because it would allow communications with the assembly machine controller (which was already fitted with this interface) to be easily established. The occurrence of a process error could then trigger a signal to stop the assembly machine, or to control the disposition of the product which could be diverted to a reject parts tray.

For reasons of cost, environment and flexibility the front-end processor hardware selected was an STE bus Eurocard computer system, together with a strain gauge amplifier constructed on a Eurocard to fit within the unit. Everything was housed in a rack which could be integrated into the assembly machine controller. The system consisted of:

1. processor board – Z80 board running at 4 MHz;
2. I/O board – IEEE-488 board;
3. A/D converter board – 8-bit conversion in 10 μs;
4. system watchdog board;
5. strain gauge amplifier.

All these boards were relatively inexpensive and readily available from electronics and computer hardware suppliers. Figure 9.3 is a diagram of the front-end processor together with the other elements of the system.

9.5.2 *Programming the front-end processor*

To ensure that the flexibility of the assembly machine was matched by the quality monitoring equipment, the front-end processor had to be capable of rapid resetting from the supervisory micro with 'master data' referring to the standard insertion force signature of each connector to be assembled. Then, for each assembly operation, the following procedure was followed:

1. Amplify the analogue strain gauge output from the pin insertion mechanism.
2. Sample and digitize the output from the amplifier and place this 'force signature' into an area of memory.
3. Compare the resultant values with the standard 'master data' values for that assembly operation.
4. Output an error signal to the supervisory micro should the

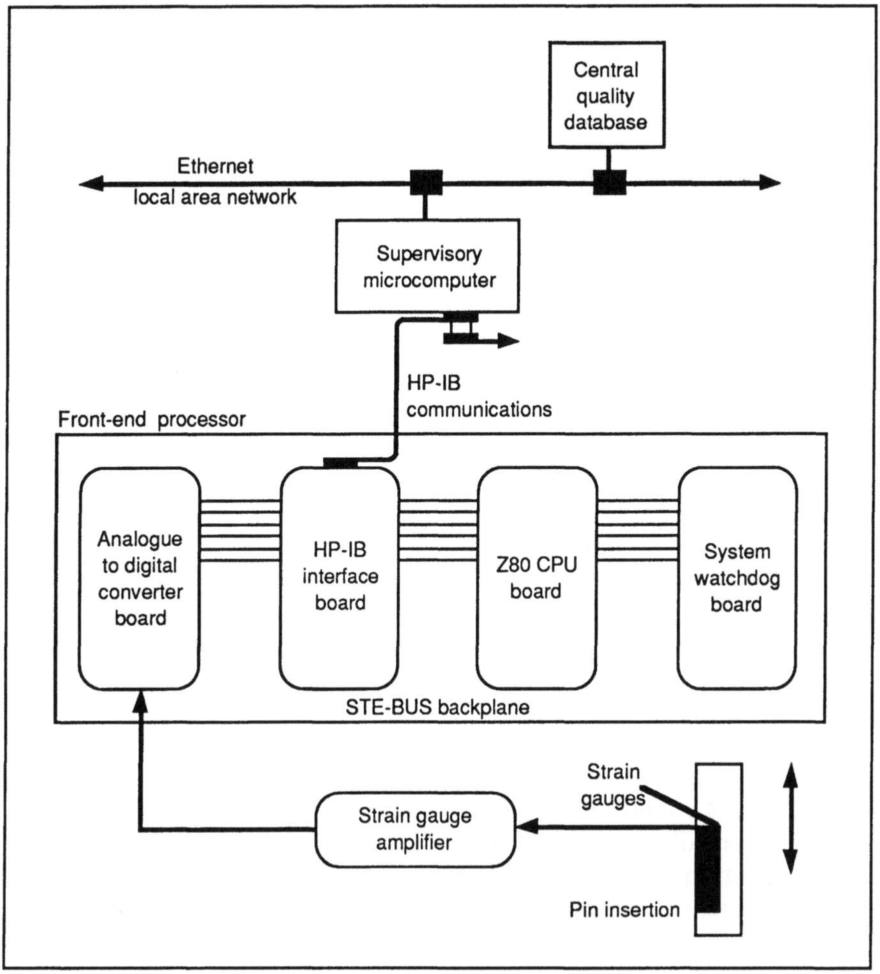

Figure 9.3 Assembly force front-end processor

comparison show a potential quality problem.

5. Output batch status information to the supervisory micro as required.
6. Self test and indicate system status on demand.

A multi-tasking facility was needed to perform these various functions. Interrupt-based task switching was used, with the interrupts occurring every 2.05ms, chosen according to the speed requirement of the assembly process. (A time of approximately 0.5ms was necessary to switch between tasks.) The processor overhead was minimized by the use of a multitasking executive to oversee the task scheduling. Dual priority time slicing techniques were used (for a practical introduction to these techniques, see Walls, 1986).

For this application, to ensure processing speed was adequate, the front-end processor firmware was written in Z80 Assembler, although for slower processes a BASIC program might have been adequate, and much more easily developed. A specialized development system with in-circuit emulation facilities was used for program development, and the completed program was then transferred to an EPROM to run the front-end processor. System functions are explained in more detail below.

With regard to communications with supervisory microcomputer, the IEEE-488 interface board was initialized on power-up or reset to await communications from the supervisory micro. A simple communications protocol was established. The supervisor would send a message, and a defined response was returned from the force monitor system to indicate correct system status. Following a satisfactory response, the supervisor would download 'master data', which was stored in RAM, for the particular insertion task to be carried out.

On product change new master data would be loaded by interrupting the front-end processor from the IEEE-488 interface. For quality problem notification the front-end processor I/O board could request attention on the interface, and the supervisory micro as interface controller would then conduct a poll to determine the source of the attention request and accept the message, which consisted of a short text error message. As an alternative, the front-end processor could be instructed at any time by the supervisory micro to return the actual force signature data relating to the defective insertion, rather than the error message. This facility was included to help in the identification and rectification of error conditions.

In addition, if the current batch quality status was required the supervisory micro could interrupt the front-end processor to obtain the latest figures of good and bad pin insertions, maintained continuously from batch commencement.

The task of data collection consisted of collecting data from a single sample of the A/D converter output (digitized from the strain gauge amplifier) and loading it to an incremented RAM address, forming a set of values defining the actual insertion force signature for the current pin insertion. This task was carried out by the interrupt service routine which was also responsible for task switching. There were important detailed considerations of timing to be addressed in the design of this task. The two sets of data had to be effectively aligned on the time axis so that a meaningful comparison could be made.

In the first prototype, data items would only be loaded to memory if the level of the signal input exceeded a preset trigger level, indicating that an insertion had started. Triggering caused a lot of trouble during system development, as the stability of the analogue electronics in the strain gauge amplifier was poor and subject to drift. This could result in data collection from spurious insertions or failure to detect when an actual insertion occurred. The firmware was modified to eliminate the problem by dynamically updating the trigger level.

As regards data analysis and error monitoring, after data collection from each insertion, the force signature was analysed to determine whether it indicated a process malfunction. The signature analysis method which made use of the largest amount of process information, and hence offered the most potential for automatic error recognition in future developments, was chosen – this being the real-time envelope evaluation of the signature waveform. The data analysis task compared the set of collected data with the master data, using an evaluation based on an envelope surrounding the master signature. The envelope consisted of a band centred on the master signature values, the width of the band or tolerance determining the sensitivity of the analysis. The occurrences of values outside the envelope were counted, and an error flag set if more than a certain number of excursions had been detected. The width of the envelope could be altered in software by the supervisory micro at any time in the monitoring process, by interrupting the front-end processor via the IEEE-488 interface. This facility allowed the sensitivity of the front-end processor to be changed to accommodate different products and process conditions.

System status checking was considered essential to ensure that the process was monitored with sufficient reliability. The system watchdog board would interrupt the processor board on the occurrence of a power fluctuation or STE bus time-out. In addition, as a background task, the processor was required to write to the watchdog board every second to indicate that it was functioning normally, or a system reset would be generated by the watchdog board. The front-end processor system status was displayed continuously on an alphanumeric LCD display.

9.5.3 *Supervisory microcomputer software*

Two programs were developed for the supervisory micro, the first for initial communication test purposes, which allowed selection of communications functions from a menu. The communications were manually initiated. This program proved invaluable for development and debugging of the front-end processor firmware.

When the basic communications had been established, a second, more sophisticated, program was developed for process monitoring. This program automated the tasks of obtaining the master signature via a LAN from the CQDB computer, sending it to the front-end processor, and then monitoring the interface for error messages. A standard batch end report was sent to the quality database at the end of each batch. The process operator could instruct the supervisory micro to perform the following tasks, from a menu:

1. interrogate the front-end processor for current batch status, returning the number of 'good' and 'bad' insertions;
2. display the force signature for the last 'bad' insertion;
3. change the sensitivity of the force signature analysis by altering the size of the tolerance envelope;
4. reinitialize the front-end processor;
5. load master data, obtained from the CQDB, for a new product;
6. send a summarized batch-end quality report to the CQDB.

The system could automatically assess the process status from the frequency of error messages received, and instruct the assembly machine control computer to shut down the process if necessary.

The supervisory micro software was written in C in order to integrate the force monitoring communications and active networked communications with the quality database. The software incorporated a sophisticated windowed user interface showing details of communication and process status, and was written in such a way as to allow easy configuration of the supervisory micro for multiple front-end processors with different functions and communications requirements. This approach was adopted to enable one supervisory micro to monitor a number of processes (or process stages) via different front-end processors, which could include product inspection as well as process monitoring systems. In this case a vision system application was also linked to the system to conduct post-process inspection activities.

9.6 INTEGRATION WITH THE IQS

In-process quality control is an important primary source of quality data. At present it is often collected for process control purposes, and not as an indicator of product quality within the quality system. This is a waste of valuable information. With relatively minor extra investment, many automated process monitoring and in-process QC systems may be adapted to supply data to an IQS.

As an example of full integration with the IQS, the front-end processor described above had the ability to accept new process parameter data, in order to monitor different products, from the local quality data collection (or 'supervisory') micro. This computer was also able to communicate via a LAN with the quality database, situated on a remote workstation, which maintained force signature process specifications for each process. Hence the flexibility of manufacturing was not restricted by the process quality monitoring activities.

The front-end processor reported simple error data, which was collated into a summary report for the database, and despatched at the end of each batch. A simple form of automatic process control integration was also available at manufacturing cell level.

9.7 CONCLUSIONS

While the best solution to manufacturing quality problems is effective process improvement to eliminate variability and hence the possibility of defective production, this ideal is not always practical. In-process QC is the best alternative. Effective and reliable in-process QC can (in theory) banish non-conformance, allowing the elimination of post-process inspection activities.

In-process QC can be exercised on the basis of process parameter monitoring or in-process inspection. The choice between the two methods depends on the characteristics of the particular process, and on the extent to which the relationship between process parameter data and product quality is understood.

Automation in the collection and analysis of in-process quality data has an important part to play in increasing the speed and effectiveness of process control. This is a vital area of functional integration for the IQS. Damaging trends in the process may often be detected early enough to enable process adjustment before any defectives are produced. The data can also be of great value to quality improvement activities.

Process settings and parameters should be considered as a vital part of the quality information in a company. An increasing number of processes are controlled by computers which can accept this information from a central source. Holding this vital information in a quality database, and downloading it as necessary, should allow a reduction in change-over times and greater control over initial process performance.

10

Coordinate measuring machines

10.1 INTRODUCTION

CMMs are flexible dimensional measuring devices, widely used for the post-process inspection of machined, moulded and fabricated components. CMMs operate with the aid of three perpendicular axis measuring scales and (on the majority of modern CMMs) a special contact-sensitive probe to detect the surface of the component. Most CMMs are used as contact measuring instruments, but some systems are available which use a non-contact principle, with a laser transducer unit or machine vision camera instead of the touch trigger probe.

When a component is to be inspected, it is positioned on the CMM table, and the contacting stylus of the probe is moved to touch the appropriate feature (perhaps a machined face or the perimeter of a hole) of the part. This is performed by moving the axis slides to which are attached the measuring scales. The scales are read at the point of probe contact to determine the *X*, *Y* and *Z* coordinates of the position in question with respect to some reference point in the instrument working space.

The level of CMM automation varies widely. The most sophisticated machines are capable of fully automated inspection and reporting of results. The computerized CMM can analyse the component dimensions to determine out-of-tolerance conditions, and use statistical techniques to determine capability and trends. More than half the CMMs sold are now fitted with some level of computer control.

The market for CMMs world-wide is expanding, because of the increasing emphasis on quality in manufacturing, and because of a shortage of skilled inspection personnel able to use traditional metrology instruments effectively. Speed and convenience in measurement, a reduced need for inspection fixtures and gauges, and a relative freedom from error are the genuine advantages of the CMM. However, there is also an image factor involved in CMM purchase, especially for the smaller manufacturer who may hope to impress potential customers with his high-tech inspection facilities.

At the time of writing, an estimated 350 machines were being installed world-wide each month, with perhaps half of these manufactured in Japan and much of the remainder in the USA. A number of significant European manufacturers also exist. The market is highly competitive, and prices have declined in real terms.

Most CMMs are sold for general shop-floor use, but there is a smaller market for true high-precision instruments designed to be used in standards rooms under temperature-controlled conditions.

In addition to quality inspection, CMMs have application as digitizing devices for 'reverse engineering' where the geometry of a component (perhaps a packaging item such as a yoghurt tub) may be measured and represented to a CADCAM system, which can then be used to program a CNC machine tool to produce a mould tool, hence manufacturing a replica.

10.2 CMM CONSTRUCTION

The CMM is a deceptively simple device to manufacture, with real precision required for a relatively small number of components, such as the surfaces of the measurement table and sliding elements. However, CMMs can be very expensive to purchase, as expensive as sophisticated CNC machine tools.

CMMs vary in size from manually operated bench-top machines (with a working volume of perhaps $0.05\,m^3$) to machines large enough to inspect a motor vehicle body. The measuring envelope of the CMM is defined by the largest X, Y and Z movements of the probe, but in practice is also affected by the size of the worktable, and access for components of various shapes. In order to cater for different shapes and sizes of components, most CMM manufacturers offer a range of instruments with different configurations.

CMMs are usually constructed using a granite surface plate as a base. This forms the flat surface on which components are placed for inspection. The rock slab has been relieved of its stresses over millions of years, and can also provide a stable foundation for guideways which will take the superstructure of the CMM.

There are a number of basic CMM configurations, which are described in standards such as BS 6808:1987, Part 1, and illustrated in Figure 10.1. The most widely used is the bridge type where a moving bridge supported on two upright columns suspends the vertical quill above the surface table. The gantry type is similar. Cantilever machines, which use a single vertical column and horizontal arm, are also popular. A number of other possible configurations are rarely manufactured at present.

Bridge and gantry configurations offer greater rigidity, are better able to resist deflections due to gravity acting on the moving elements of the machine, and hence are the most popular types. The high inertia inherent in bridge and gantry designs has the disadvantage of reducing the practical slide accelerations. Lengthy inspection times are now recognized as a problem for CMMs, and cantilever configurations have the potential for lower inertia and hence shorter measuring times. Bridge and gantry CMMs also have the disadvantage of requiring a large overhead clearance, and restricting access for automatic loading of components.

Granite is often used for the moving elements, such as the bridge beam, as well as the base. Ceramic beams and quills, lapped to ensure accurate straightness and flatness, are also used in CMMs for their hardness, good coefficient of expansion, stability and low inertia. Other machines use metal, often aluminium, elements.

Smooth motion of the slides is achieved using air bearings. As well as minimizing sliding friction, these also have an averaging effect to reduce the consequences of surface imperfections on the slideways. A suitable air supply is required.

Movement of the machine axes is entirely manual on lower-priced CMMs. Most manual machines are fitted with handwheels for precise movement, although many users prefer to move the slides by direct hand pressure.

More expensive machines are fitted with motor-powered axis drives, using DC servo motors operating through special mechanisms which enable each axis drive to be readily disconnected by a switch, allowing manual movement.

The axis drive motors are usually controlled using a joystick. Even those instruments fitted with drive motors are still sometimes positioned by hand, as inspectors often prefer the more direct control they believe is possible with manual operation.

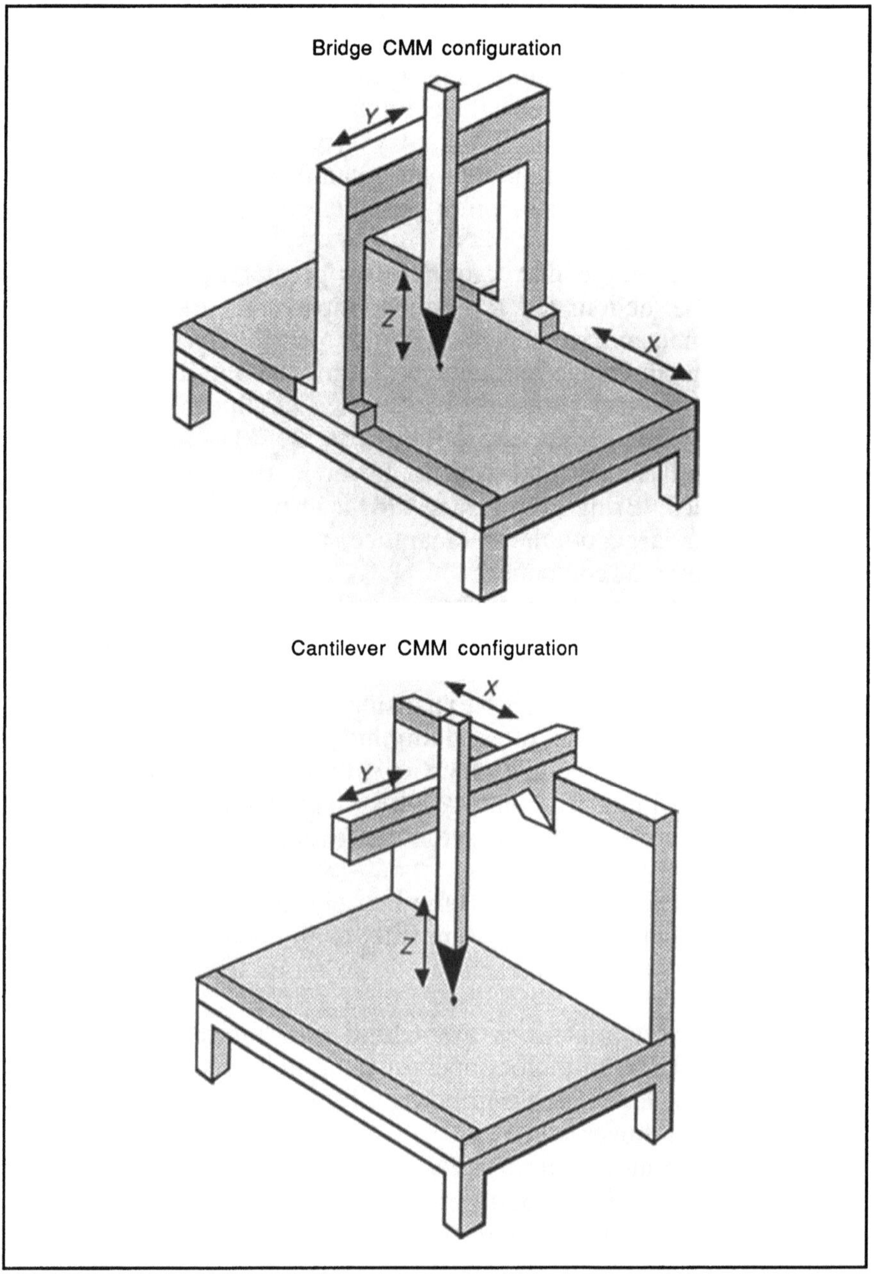

Figure 10.1 Typical CMM configurations

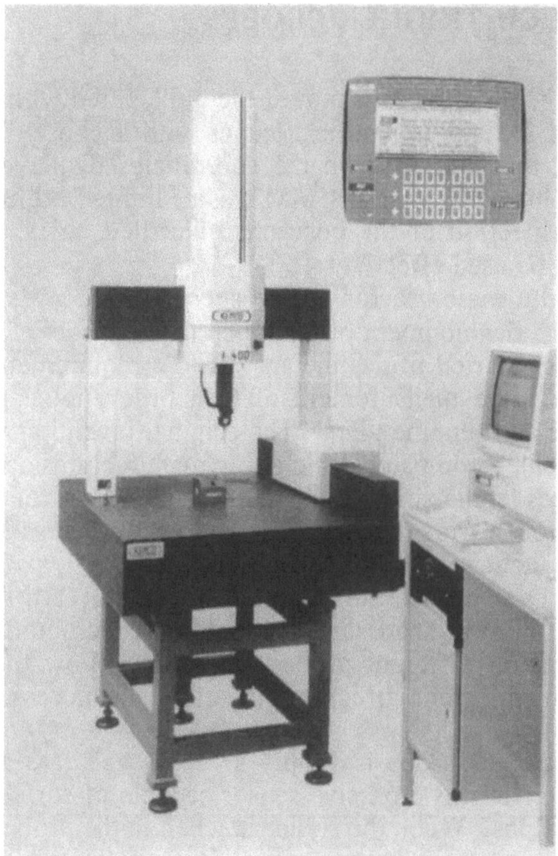

Figure 10.2 A computerized CMM
(reproduced with permission of Keeling Metrology Company Ltd.)

CNC machines are fitted with axis drive motors controlled by the computer, which may be programmed to move the probe through a sequence of measuring positions automatically. Such machines are also referred to as direct computer control (DCC) CMMs. Figure 10.2 shows such a machine manufactured by the Keeling Metrology Company.

10.3 TOUCH TRIGGER PROBES

On the first CMMs the user moved the quill, fitted with a simple rigid stylus, until the stylus was in contact with a chosen feature of the component, then noted down the coordinates displayed on the axis readouts. This early type of CMM was of limited usefulness, but was ideal for inspection of the positions of drilled holes, where a taper probe could be used effectively.

Perhaps the major factor in the success of CMMs has been the invention and development of the touch trigger probe. This brilliantly simple device started life as the answer to a requirement for accurate, low-contact force measurement of Concorde engine pipework at a Rolls-Royce aero-engine plant. The original invention was developed by the Renishaw company, which now manufactures a range of touch trigger probes, some suitable for CMMs and others for CNC machine tools. The touch trigger probe is now almost universally specified by CMM manufacturers.

Touch trigger probes are mounted on to the CMM quill using a probe head. Probe heads are available to accommodate either one probe or several different probes, and some allow the probe to be orientated in different directions to attain the necessary inspection angle.

A touch trigger probe is simply a highly accurate and repeatable switch, which acts as a passive sensor capable of surface detection in three dimensions. When the stylus attached to the probe is moved into contact with a component surface, it is deflected against spring pressure, and one of a number of electrical contacts is broken. The change in resistance as the contact begins to open is detected and used to inform the system that a surface has been encountered. CMM probes are available with repeatabilities of 0.25-1.0 μm depending on the type. This figure refers to twice the standard deviation obtained from ten measurements in the same direction. A small and repeatable amount of pre-travel precedes the triggering, and this is automatically compensated, for each stylus length, during stylus tip calibration.

There are a number of different types of stylus which can be attached to a touch trigger probe, depending on the type of feature which must be inspected. Standard styli come in three categories: the most common is the single ruby ball on a stem, used for general inspection. The industrial ruby tip is highly spherical and wear-resistant. Star styli incorporate four or five ruby balls on stems for the inspection of complex components. Disk styli are used to inspect

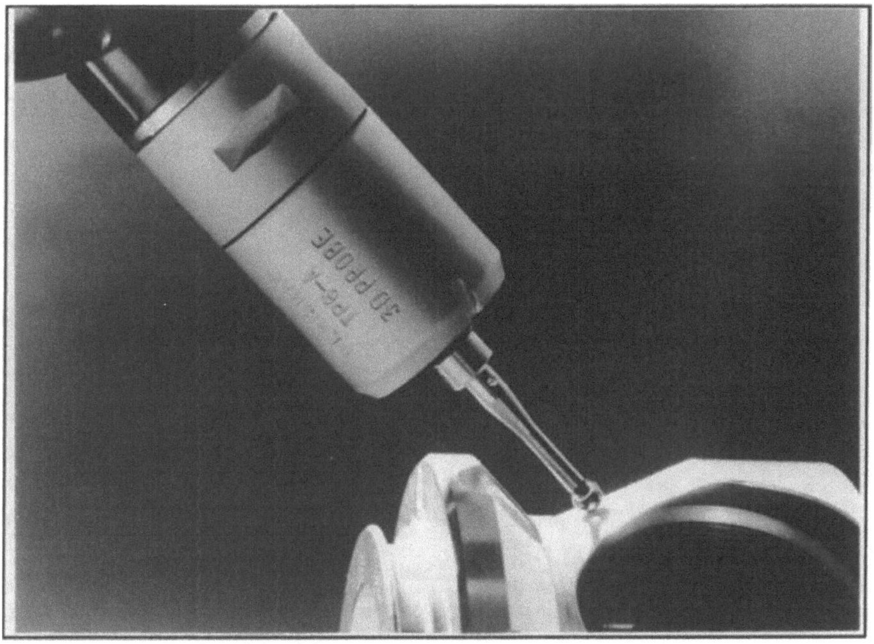

Figure 10.3 A touch trigger probe
(reproduced with permission of Renishaw International Ltd.)

undercuts and holes. Special styli are available for tasks such as screw thread measurement.

Figure 10.3 shows a Renishaw touch trigger probe fitted with a standard single ruby ball stylus, as it is brought into contact with a component surface. This slim type of probe is useful for probing in restricted spaces. It is capable of triggering in both positive and negative probe axis directions, which means that it is also suitable for use with a disk stylus.

The touch trigger probe provides a signal to allow the 'latching' of the CMM coordinates at the moment of probe triggering. Probe signals are converted by an interface into a form acceptable to the CMM readout unit or controller. On most modern CMMs the inspection data are recorded with the aid of the touch trigger probe and a computer, which captures the coordinate information and calculates the component dimensions.

Substantial stylus over-travel is possible before damage to the probe or component will occur, giving the operator or controlling computer time to stop the motion of the slides. The stylus contact forces are

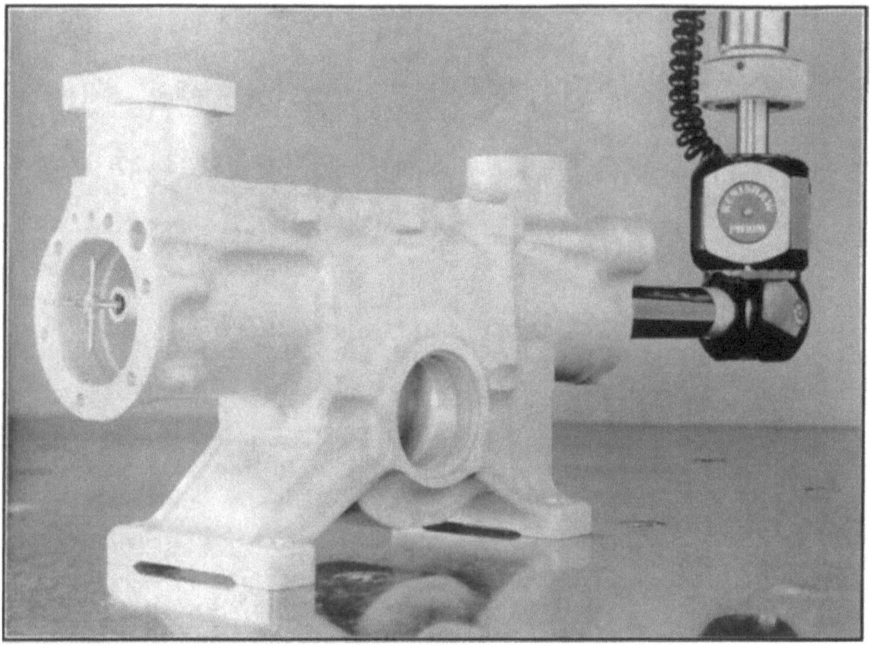

Figure 10.4 A motorised probe head with extension bar
(reproduced with permission of Renishaw International Ltd.)

very low, of the order of 5-10 g, so that deflection of the workpiece or the CMM is unlikely to be significant.

Motorized articulated probe heads increase the versatility of the CMM by allowing automatic orientation of the probe in the direction required for a particular measurement. Figure 10.4 shows a motorized probe head fitted with an extension bar to measure inside a deep bore feature. A five-ball star stylus has been fitted to the probe for this application.

10.4 CMM USE AND PROGRAMMING

At start-up time, every CMM and probe combination must be 'datumed'. This involves installing the probe which is to be used for inspection, and following a routine including the touching of each stylus of the probe on to a calibrated object, usually a sphere, which is attached to the table. By various methods, the centre of this sphere is set to be the origin (zero point) of the *X-Y-Z* axis coordinate system. Usually this procedure must be repeated whenever a different probe is to be used. An aid to flexibility is an automatic probe exchange system, allowing a number of different probes, probably with different styli, to be used in sequence without redatuming. The probe head shown in Figure 10.4 has the facility for automatic probe exchange, with repeatable kinematic location for each probe or extension bar.

Next, the component for inspection will be placed on the CMM table. It may be located in some definite position determined by a fixture bolted to the table, or it may be simply placed in approximate position by eye. In the latter case the position and alignment of the part must be determined as a first step in the inspection program for that part.

Manual or joystick-controlled machines can be used much like any other inspection instrument, in that they are user-controlled. The sequence of feature measurements may be left to the skilled inspector, using only component drawings, or it may be defined formally as a part inspection plan or program which is built up for each component to be inspected. The part inspection program may be recorded on an inspection sheet with manual entry and analysis of the dimensional data, or by using the CMM computer with suitable software.

If the program is defined using the CMM computer, the sequence of measurement elements must obviously be strictly followed for each subsequent part, as otherwise the coordinate data obtained will not match those expected by the computer. Instructions must be provided on the computer screen to guide the user through the sequence. The CMM used in this manner acts as a CAI station.

Effective software to ease the inspection part-programming process is an important area for CMM users, and for CMM manufacturers strong competitive pressures have led to rapid developments.

The ideal CMM programming method would involve the use of geometric data obtained from a CAD system to generate the inspection program directly. This is an important area for integration in a CIM system. The requirements are similar to those for the automation of process planning (and therefore part program generation) for machined parts. The features (such as bosses, slots, holes and shoulders) of the component geometry must be recognized by the computer from the CAD data.

Unfortunately CAD data are not provided in terms of features, but as lower-level descriptions such as primitive solids, vertices, faces and edges. Several different methods have been developed to recognize features from CAD data (both 2D and 3D systems), and research continues to develop automated process planning systems. When effective automated process planning systems are perfected, CMM programming systems will not be far behind. Anticipating the general integration of design and dimensional measurement, a standard specification for the transfer of data between CADCAM systems and CMMs has been agreed in the USA, called the Dimensional Measuring Interface Specification (DMIS). Software support for the DMIS is now available from a number of sources.

A useful degree of computer assistance is possible without expecting the computer to understand component features. Trained humans have excellent spatial reasoning abilities, and are easily able to recognize features represented on a technical drawing or sample part. The CMM programmer therefore selects the type of feature from a menu of standard functions, which includes such elements as circles, cylinders, planes, lines and spheres, which may be connected or intersected, and on which calculations, such as those required for geometric tolerances, may be performed. The programmer also determines the sequence of measurements and output (for example, the printout or SPC chart) required, and enters operator instructions.

The user interface may include 'touch-and-learn' or graphics abilities to allow selection of component features to be inspected; the software can then help string them together into the inspection sequence with minimum user involvement.

Programming the CNC CMM to inspect a new component is potentially a time-consuming business, adding to the complexity of the programming task. One or more powered axis moves will be required for each probing point and, if the programming is poorly performed, expensive damage could well result.

CNC machines are provided with a more extensive library of procedures to ease the programming task by automating the commonly used set-up and measurement activities. The programmer can select these 'macros' from the library, and incorporate them into the inspection part program with a few key stokes. Typical macros allow datum setting, automated part alignment, measurement of circle diameters, communication with external computers and the selection of inspection reporting options.

10.5 CMM UNCERTAINTY AND ERRORS

All CMMs suffer from a complex mix of errors which cause uncertainty of measurement. Errors can be systematic or random. Only systematic errors may be predicted and compensated for. Systematic 'hardware' errors result from the geometrical deficiencies and distortions of the CMM construction, and from measuring scale ruling or zeroing errors. Probing errors such as variation in contact pressure and frictional effects tend to be random.

The uncertainty of any individual measurement made with a CMM is difficult to specify, because it depends on the error vectors (the magnitude and direction of the difference between the true position and the position indicated) of each measuring point involved. There is uncertainty in every measurement carried out using a CMM, as with every inspection instrument. The degree of uncertainty is defined by the limits of error, and hence the uncertainty is closely related to the more commonly understood idea of accuracy, which is similarly defined.

To complicate matters, there are different types of uncertainty. BS 6808:Part 2 (1987) defines CMM performance in terms of uncertainty in measurement of length, three-dimensional measurement of length, and also volumetric accuracy. Volumetric accuracy refers to the performance of the complete measuring system when carrying out a compound multiaxis movement. Three-dimensional uncertainty is a special case of volumetric accuracy, defined as the difference of the measured and true lengths between two opposite points on parallel faces of an artefact when approached from opposite directions. According to BS 6808 the three-dimensional uncertainty should be specified by the CMM manufacturer using a formula which takes into account the measuring length and certain special constants. BS 6808:Part 3 was published in 1989 as a Code of Practice and guide to the use of Part 2.

It must be appreciated that uncertainty (or accuracy) is not the same as repeatability of measurement. Repeatability is no more than the closeness of agreement between the results of successive identical measurements carried out under closely controlled conditions. Repeatability is an indication of the extent of the random errors in a measuring system, and ignores the systematic errors. Hence an instrument may be correctly specified as being repeatable to well within the required measurement performance, but its uncertainty of measurement may be substantially worse.

Another, rather misleading, criterion of performance which is sometimes quoted is measuring resolution. This is much more easily quantified than the uncertainty, simply by reference to the axis measuring system specifications. Most CMMs use Moire fringe methods with precision ruled linear scales, and have typical resolutions of 0.5 or 1 µm. It may be confidently assumed that the measuring resolution will be substantially smaller than the repeatability, which will in turn be much better than the uncertainty of any measurement.

Errors can also arise from programming and rounding errors in the control software. It is revealing to compare the results obtained by different manufacturers' CMMs when measuring the same object, such as a sphere diameter calculated from touching three points on the circumference. Assuming no actual software bugs exist, different techniques of calculation, combined with the differing numbers of significant digits handled by the programming language in use, can have a substantial impact on such measurements. Simple measurements, such as those of length in one axis, are not so seriously affected by these problems.

For example, the worst case volumetric accuracy achieved by one CMM and probe combination, when measuring a 15 mm diameter sphere in a recent extensive test, was no better than –0.031 mm with a 95% confidence interval of ±0.0104 mm. This poor performance is probably not typical, but it should give CMM users food for thought. More than one engineering company known to the author is experiencing difficulty in ensuring that its CMMs are capable of a sufficiently good measuring performance accuracy to inspect precision components under shop-floor conditions.

Complete systematic error compensation in software (using an error map of the complete working volume) is used by many manufacturers, although such software can itself introduce rounding and interpolation errors.

In addition to the geometrical, software, probe and measuring scale errors inherent in the CMM system, the most important consideration as regards uncertainty of measurement is temperature. Thermal expansion of the various elements of the CMM will result in inspection errors.

On the widely used bridge-type CMM, the horizontal beam of the bridge is supported at either end by vertical columns which travel along the machine base. The material used for these columns is potentially important to the measuring performance of machines which are not used under temperature-controlled conditions, due to thermal expansion. Cast iron, fabricated steel construction or aluminium castings are all commonly used; the last material expands most.

The measuring temperature maintained in standards rooms is 20°C, and this is the temperature at which performance results are usually specified by CMM manufacturers. The specified levels of performance will not usually be achievable under shop-floor conditions of variable temperature, which can introduce thermal gradients causing distortion of the CMM structure. Machines constructed from elements with a low thermal mass will track the shop-floor temperature changes more rapidly. In this respect lightly constructed cantilever CMMs have an advantage over massive granite bridge-type machines. BS 6808 includes a simple test to take account of the effect of temperature. Manufacturers of some CMMs (especially those incorporating high-expansion-rate aluminium elements) have started to introduce automatic temperature compensation in software.

Vibration, especially for CMMs sited close to large machine tools such as power presses, can also affect measuring performance, as can air and electricity supply problems.

10.6 PERFORMANCE TESTING OF CMMs

Performance testing of CMMs falls into three categories: the initial qualification by the manufacturer; periodic full recalibration; and periodic verification.

Full calibration of CMMs is carried out for acceptance testing, to maintain traceability to international dimensional standards, and to achieve error compensation in software. Verification is carried out as a relatively rapid assessment of the working accuracy, usually only valid in a proportion of the working volume, but which can be used to determine if any changes have occurred in the machine performance.

There are two basic methods of CMM calibration: using a laser interferometer or a precision-calibrated artefact. Various national and international standards for both methods exist. Unfortunately, there is often little association between the two types of standard, so it is difficult to relate measurements taken using the different methods.

It is, of course, still possible (though time-consuming and prone to error) to calibrate a CMM using the classic methods of metrology, such as the straight edge, electronic level and square.

The latest standards, such as BS 6808:Part 2, require far more measurements than were usually undertaken previously, which can be very time-consuming. A full calibration must measure the departures from the ideal for all the 21 degrees of freedom of a three-axis CMM. It may take up to two days to perform, and will be required every three to six months to maintain traceability.

Volumetric calibration by laser interferometer is carried out by measuring displacement errors along a number of lines in the machine working zone. Measuring the displacement errors along two parallel lines for each axis allows pitch and yaw to be determined. Squareness and straightness error require the measurement of two diagonals and one additional line, while roll errors can be derived from straightness errors on two parallel planes. Formulae for error calculation from these measurements are well documented, and are incorporated in calibration software.

Full calibration using an artefact can be very time-consuming, with a large number of repetitive measurements being required. BS6808 suggests the use of one-dimensional artefacts such as bidirectional step gauges and end-bars (or long gauge blocks). The performance of the CMM is determined by measuring the artefact in various different orientations, then assessing the three-dimensional uncertainty in length measurement, and hence the volumetric accuracy.

In other countries different artefacts are used. In Germany, for example, two-dimensional artefacts are used including the expensive Zerodur bore plates and various ball plates, where ceramic spheres are inlaid into a steel plate in known locations. Naturally such artefacts also require periodic calibration, probably using a laser interferometer. Ball plates are not considered effective for the calibration of the most accurate laboratory CMMs.

The standard also recommends daily performance verification, which is often performed using calibrated workpieces. Purpose-made calibrated workpieces can give an excellent indication of the accuracy of the machine in connection with the type of object it must inspect.

An alternative to the use of a calibrated workpiece or other static artefact for periodic verification is a mobile artefact. A typical example of a mobile artefact is the Renishaw machine checking gauge. This uses a well-balanced arm of a fixed length pivoting freely on a fixed spherical bearing. Projecting radially from the other end of the arm are two parallel rods between which the CMM probe stylus is placed. The stylus is then moved over a spherical path taking the arm with it, position readings being taken at various points. The readings are taken by moving the stylus towards the pivot, until it reaches a precision ball set in the end of the arm. For a perfect CMM, all the measured radii will be identical. The volumetric accuracy is simply determined by subtracting the smallest from the largest reading obtained. Repeatability is determined by comparing the equivalent readings taken during different verification checks.

Unlike a full calibration, verification with a mobile artefact does not cover all of the working volume to discover the sources of error and calculate the positioning error at a particular point. However, the time taken is only 15-30 minutes, meaning that the CMM may be verified every day before use for inspection. There is no comparison to traceable standards, of course, unless the artefact has been calibrated, which is not the case with existing mobile artefacts.

10.7 TRAINING

A problem with the CMM is that many users do not really understand how it works; it is a black box which produces the answers. The user of traditional metrology equipment must understand the principles, as well as how to use the instruments, or he can obtain no valid results at all.

Management must ensure that CMM users, especially those using CNC machines, are fully aware of the principles and limitations as well as the operating procedures. CMM manufacturers run training courses when they sell machines, and some metrology courses are available from colleges and universities. Various commercial training organizations also advertise suitable courses.

It must never be forgotten that the whole edifice of modern manufacturing rests on the foundation of component interchangeability, which is itself dependent on the ability to measure accurately. Even if the skills to use outmoded manual measuring instruments disappear, the knowledge of the fundamentals of metrology, such as traceability and the causes of errors, must be retained.

There is nowadays a world-wide network of national standards organizations, coordinated through the ISO, to ensure that measurements throughout the world are made against the same ultimate standards in such matters as length, weight and electrical units. National organizations organize calibration facilities to ensure traceability of all measuring equipment to international standards. All companies which carry out precision dimensional inspection using

CMMs (or any other method) should ensure that they have the skills and knowledge to make use of these services.

10.8 INTEGRATION WITH THE IQS

CMMs can be used as either CAI workstations or fully automated quality data collection stations, depending on the abilities of the instrument.

Manually operated CMMs – either with or without motor driven axis movements – which use a computer to collect and analyse the dimensional information from the inspection task are ideally suited to the role of a CAI workstation, provided the computer can be programmed to communicate with a remote quality database.

CNC machines can also take the role of fully automated post-process inspection stations. They can operate in both traditional and automated manufacturing environments such as flexible manufacturing systems. In an FMS, a CMM can provide process control information directly to the FMS controller or an individual machine tool, and also send summarized quality reports to the CQDB.

CMMs operating as either CAI workstations or automated inspection stations can obtain their inspection programs from a quality database, which may in turn have acquired the information from a suitable CADCAM system.

The computer communications requirements for vertical data integration within the IQS are not, unfortunately, fulfilled by all current CMM controllers. The ability to send and receive data via an LAN requires suitable communications software, and often special hardware as well. The communications software may not be capable of working directly with the CNC control software, with the consequence that the user will have to save the inspection results to a disk file, terminate the CMM control program and start up the communications software, then establish contact with the remote database machine and initiate a manual file transfer.

This is highly inconvenient and wastes inspection time. The best answer is the provision, by CMM manufacturers, of integrated communications facilities able to communicate with suitable industry standard computer networks. An alternative is the use of CMM control computers capable of multitasking operation, perhaps using the Unix operating system. Those CMM manufacturers who fail to provide a suitable answer to data integration will find that their competitive position is adversely effected.

10.9 CONCLUSIONS

In many manufacturing companies the CMM has largely taken over from traditional metrology equipment, for all but the simplest dimensional measurement using hand-held instruments. Instruments like comparators, toolmakers' microscopes, squares, and taper plug gauges have often been laid aside, and the skills to use them may not survive long. However, metrologists have over the years seen techniques for the measurement of many types of feature, such as roundness and surface finish, taken over by automated methods, and no discernible harm has come of this.

The main advantages of the CMM are convenience of use and flexibility; most shapes can be inspected with appropriate probes, and the opportunity for human error is minimized. In addition, the requirement for expensive inspection gauges and fixtures can be greatly reduced.

Although greatly simplified by the ease of use of the CMM, inspection of complex components remains a lengthy task, because of the large number of moves and computations required. This is true for CNC machines as well as for manually controlled CMMs.

The CMM is not ideal for the inspection of every type of feature. For example, certain internal details may require the use of traditional instruments if no probe can reach effectively.

The most important reservation concerns CMM measuring uncertainty. It is the experience of many manufacturers that the actual measuring performance of machines situated on the shop floor is only marginally adequate for the inspection of precision components. This is true even for the most carefully maintained and calibrated machines.

Despite the existence of quality systems designed to ensure adherence to calibration schedules, there are far too many companies where slackness has crept in, and where CMMs are not properly looked after. It is essential that calibration and verification procedures are carried out effectively to ensure that the quality data obtained from inspection are reliable.

The quality information produced by a computerized CMM is ideally suited for use within an IQS as it can be communicated directly to a quality database, the CMM acting as a local quality data collection node. Unfortunately, many CMMs are not yet provided with suitable communications software.

For process control purposes, CMMs are ideally suited for use in an automated manufacturing environment, such as an FMS, for first-off inspection and subsequent process control of CNC machine tools.

11
Machine vision

11.1 INTRODUCTION

Machine vision is an important technology widely used for non-contact inspection. It involves the use of a computer fitted with specialized hardware for the acquisition of a visual image. Once the image is captured, the computer will use special software to process the data contained in the image. This can provide information for tasks such as measurement, detection and recognition of objects or patterns, and many other tasks of value in QC.

Machine vision technology is well in advance of most industrial inspection applications. There are aerospace, medical and military applications of great sophistication. Inspection applications have been limited by the need to demonstrate cost effectiveness, and by the need for real-time operation in demanding manufacturing environments. As the real cost of computing power falls, the integration of vision into manufacturing systems is an area where considerable advances can be expected in the near future.

Around 80% of industrial vision systems sold are employed primarily on inspection tasks, the others mostly for robot guidance, particularly for welding. Sorting and selection of products is often integrated with the inspection function.

A number of books are available for those interested in the details of machine vision, but an outline of the technology is given below.

11.2 PROVIDING THE IMAGE – THE CAMERA AND LIGHTING

The camera will be either a standard television vidicon, or for more modern systems, a solid-state charge coupled device (CCD) or charge injection device (CID). In these cameras a lens focuses the required image on to a two-dimensional grid array of photosensitive semiconductor elements, which are affected by the incident light, and sense the image in the form of pixels, each provided by a single element of the grid.

The ability to capture the information in the image depends on the number of pixels in the grid. Typical pixel grids measure 256×256, 512×512 or 1024×1024 elements. The greater the number of pixels available, the greater the detail and, hence, the better the quality information which can be resolved by the camera. Image information cannot be added later by the vision system computer, although features may be enhanced.

For the simpler linescan vision systems a one-dimensional array (a single line) of photoelectric elements is used, and the object of study must move relative to the camera to allow an image to be built up. This approach is sometimes the most suitable for discrete products moving along a conveyor belt, or for continuous moving product such as hot steel strip and extrusions.

Interchangeable camera lenses may be used to obtain the required detail and field of view, as in conventional photographic cameras. The image focused by the lens should be as large as possible to make the best use of the available pixels. Zoom lenses are the simplest method of tailoring the image to requirements. For very small objects, extension tubes are available to space the lens away from the camera body, and allow the close focusing necessary to achieve the required field of view. The light intensity of the image is changed by adjusting the aperture of the lens, as in other cameras.

Many vision systems may be fitted with more than one camera to obtain images from different sides of the object, or to observe the same features from different angles. Sometimes this is the only way to obtain enough information to make a quality assessment.

Controlled lighting is crucial to successful machine vision applications. A few objects emit light, but most are inspected using reflected or shadowed light. The light reflected by an item under inspection depends on a number of factors, particularly texture and associated reflectivity, shadowing effects, orientation with respect to light sources and camera, and the geometry of the surfaces presented. Very few applications can make do with the illumination provided by daylight or general-purpose room lighting; in fact these light sources are usually a nuisance, giving confusing reflections. Special-purpose lighting is usually required. Diffused lighting boxes are available as a background for shadowing applications. Spotlights are used to highlight features, while ringlights around the camera lens provide a shadow-free illumination for front lighting of an object. Structured lighting, often a thin beam emitted from a slot or by a laser, allows the inspection of contour features (such as steps) by analysis of the profile of the resulting strip of reflected light. Several types of lighting may be used simultaneously to illuminate the required features.

11.3 IMAGE ACQUISITION

Once the required image has been focused on to the grid of photoelectric elements it must be captured (or 'grabbed') by the computer. The speed of image capture is important to many high-speed inspection applications. The light intensity falling on each element of the camera pixel grid will be scanned sequentially, line by line of elements, and the values obtained placed in an image buffer called a 'framestore'. Framestores (a system may have more than one) are areas of computer memory dedicated to storing the image information as it is provided from the circuits which scan the camera pixel grid.

Framestores usually contain an array of single-integer numbers referring to the 'grey level' of each pixel, that is, the relative amount of light which was detected in that part of the image. The number of grey levels which may be supported by a system varies from two (black or white, known as a 'binary system') up to 256, which is far more than the human eye can distinguish.

Framestores are incorporated as part of the hardware of purpose-built vision systems, or may be purchased as separate items, designed to operate with most common microcomputers and technical workstations. Another alternative, for those who have the electronics and computing expertise, is the construction of a vision system from readily available proprietary boards designed to fit the STE or VME bus standards.

True colour vision is not widely available in commercial vision systems at the present time, although some mono systems can provide colour graphics displays. The different colours shown on the display screen of these systems are programmed to represent different grey level bands. The lack of colour vision is not a major disadvantage for most applications, as equivalent information can usually be obtained from an analysis of grey levels in the image. Colour framestores are obtainable, however, and these can be used with a suitable colour camera to store each of the three basic colour images separately in a three-dimensional array. Processing of these images can then be used to assess intensity, hue and saturation for each part of the object to be inspected. An expensive broadcast-quality camera can be required.

11.4 IMAGE PROCESSING AND ANALYSIS

Once the image information has been transferred to the framestore, the machine vision system can start to process and analyse the data, while the system can be instructed, if necessary, to start capturing the next image immediately.

To obtain useful quality information from an image, a range of image processing and analysis procedures will be carried out, depending on the type of application. These will be performed by the computer, which is the heart of the vision system. Many standard computers, especially micros, are extremely slow at processing and analysing the information in an image, because of the very large number of complex mathematical operations which need to be performed.

With a proprietary vision system, the hardware will usually consist of a dedicated computer system based around a high-performance 32-bit microprocessor. The system will be designed for high-speed processing of the data from the framestores, and will thus be far more suitable for speed-critical inspection applications. Some speed-critical number-crunching may be achieved using special purpose co-processor hardware to enhance speed.

Vision systems must also be capable of interacting with the user, and will need a terminal and other I/O facilities. Hard disk storage will usually be provided for software and visual images, and, of course, the system will require working memory for its operating system and application programs, as well as the framestore memory.

Apart from the operating system (ideally real-time) of the vision system, which provides the environment in which inspection application programs can be developed and run, a system must include one or more computer languages with which the applications can be programmed. Most suitable compiled computer languages have been used by vision systems, particularly FORTRAN and C. Interpreted languages such as BASIC and FORTH (where the program is translated and executed line by line from the source code) are also used, although these tend to suffer from processing speed problems. Some systems have special-purpose languages or 4GL application development environments to ease the programming task.

A distinction should be made between development systems which require a full set of language software, and target systems for the shop floor which will require little in the way of a software environment. These will run only the programs developed using the development system, and may also have simplified hardware requirements.

The image processing and analysis task uses a variety of techniques, the most important fall into the categories of image simplification, feature enhancement, object recognition and quantitative measurement. In practice, most practical applications include elements of all of these categories.

It is normal to simplify the image to reduce the analysis time required. A widely used image simplification technique to reduce processing delay is segmentation (or 'windowing') of the image. Here only a small area of the total image, known to contain the relevant features, is subjected to further image processing and analysis. The image can also be simplified by conversion to binary (black and white) form at any threshold of grey level suitable for bringing out the features of interest. Any area with a grey level above a certain threshold value will become white, while anything below will become black.

Feature enhancement is carried out using 'convolutions', mathematical operations which carry out functions such as the detection and intensification of edges, corners and blobs, and the filtering of anomalous pixel detail, perhaps caused by scattered light.

Although it would be possible (using a suitable low-level language) to program all the image processing and analysis functions of a vision system from scratch, it is much more convenient in practice to use the functions in a library of software which will be provided with a commercial system, linking them together with a high-level language such as C or a special-purpose vision language. In this way the programmer can develop useful inspection programs in a fraction of the time otherwise required. The range and quality of the vision functions library is one of the most important aspects of the system.

This library contains preprogrammed functions or procedures to implement the various algorithms which have been developed for vision tasks. They perform segmentation, enhancement convolutions, filtering, and thresholding to manipulate the image. To help the system analyse and interpret the information needed to carry out the task, there are functions for connectivity analysis, pixel statistics, histograms and profiles, moments of area and template matching for object recognition.

As an example of the operations which might be required for component inspection, it could be necessary to:

1. assess component orientation, integrity and completeness by converting the image to binary form, calculating the moment of area of the resultant shape, and matching to a 'template' held in memory;
2. read and interpret a printed label using optical character recognition (OCR);
3. segment the image and measure several critical dimensions, by detecting and enhancing the edges of features and comparing the pixel count between them with stored nominal and tolerance information;
4. output a signal to an automatic sorting device according to the outcome of the inspection;
5. assess and display trends using SPC methods, and display process status;
6. report process status and quality level on demand to a manufacturing cell control computer.

A complex program would be required to perform these functions, utilizing a full range of image processing, analysis, computational and I/O techniques. There are many simple machine vision systems which do not allow such full flexibility. The cheapest systems have only a limited object recognition ability, and cannot carry out quantitative measurement as the software does not have the capability.

For complex applications, most companies will need to call upon the specialist applications engineering and programming skills of the system supplier.

11.5 MACHINE VISION APPLICATIONS

Typical machine vision abilities include linear measurement, flaw detection, shape classification and parts recognition. These can be used in a wide range of quality data collection applications, wherever a considerable degree of order can be imposed on the inspection task. Machine vision systems are not yet capable of making sense of disordered or chaotic situations, as is a human inspector. The lighting and presentation of objects for inspection must be such as to enable a suitable image to be captured for analysis.

Ford uses machine vision for vehicle instrument calibration and test; British Steel uses a system which measures the flatness of hot steel strip as it leaves the rolling mill; while Du Pont inspects a variety of electrical connectors using machine vision. Companies in the food industry have found uses for machine vision in the sorting of meat and fruit, while in pharmaceuticals, systems for checking batch and date codes on bottles are in use. Printed circuit board (PCB) inspection in the electronics industry is of increasing importance.

11.5.1 *Printed circuit board (PCB) inspection*

Before the advent of surface mount technology (SMT) methods, machine vision had for some years found limited application in the inspection of bare boards, checking the identity and orientation of assembled components, and examining for correct assembly and soldering of pins on integrated circuits.

The introduction of SMT for attaching components to the PCB has made the application of the 'bed of nails' type of automatic inspection described in Chapter 12 difficult because the new technique, combined with the reduced size of the components, results in inaccessibility of their connections.

Surface mount components do not have connector pins which pass through holes in the PCB, there to be soldered to the conductive tracks on the board. Instead, these components are held on to the board by a small amount of paste, then soldered to the top surface or to both sufaces of the board as appropriate. The electrical connections are made through tiny pads on both board and component which must be precisely positioned together. This positioning is achieved by automatic machines, the latest models being extremely accurate and reliable.

Pre-soldering inspection is rarely required, but the automatic soldering process can introduce quality problems. Typical problems involve chips which are washed away by the solder flow, and solder bridges which may be formed between adjacent conductors. Machine vision is ideal for the detection of these faults.

Special-purpose machine vision systems are available, which have cameras mounted above and below a PCB holding fixture. The fixture is moved under program control by motor drives, to position each component on the board under the cameras for inspection.

11.5.2 *Car body inspection*

Machine vision has been successfully applied to a number of automotive QC functions, many of them on-line. One interesting and unusual application which the author has examined is the 100% on-line inspection of the dimensional accuracy of Rover 800 series car bodies (fully described by Mahdevieh, 1987). Here 62 gauging modules each incorporating a laser and a CCD camera are used to illuminate and measure up to 288 gauge points on the bare body-in-white. The car body is approximately presented beneath the measuring gantry, with errors due to body orientation handled by the software.

The camera signals are multiplexed to two Automatix vision systems, which in turn report to a third master system, then via a local programmable controller to the main assembly-line controller. SPC facilities are built into the software, allowing trend detection and feedback process control. Average and range control charts, together with histograms to illustrate process capability, are available for every gauge point. The quality data are archived locally, and printed as necessary.

Calibration has been given special attention. Each laser/camera module can be calibrated off-line on the bench, while an on-line calibration fixture (itself calibrated off-line with a CMM), holds up calibration targets for the cameras.

The system can measure three body types at a rate of one body every 40 s. The system uses nominal dimensional data derived from a CAD system, and measures with an accuracy of ±0.1 mm. Rover quality personnel use the system to judge the dimensional conformance of the (largely automated) body build in crucial areas such as overall geometry and door or glass apertures.

This method of body inspection has proved to be so successful that a similar system has also been installed by Rover at the later 200 series body plant.

11.6 INTEGRATION WITH THE IQS

Machine vision systems are well suited to integration with an IQS, either as front-end processors to a local quality data acquisition computer, or as automatic inspection stations. The communication abilities of the system determine the role which is most appropriate.

The most sophisticated vision systems are highly capable general-purpose computers in their own right, running multitasking operating systems with full networking communications abilities. Such systems are capable of operating as fully automatic inspection stations, downloading inspection programs and product specifications from a database, and reporting summarized quality data.

Other systems will require to be interfaced to the IQS via a general-purpose local quality data acquisition computer with suitable networking facilities. Communication with this computer, which may be a low-cost IBM PC-type micro, can be achieved using the relatively simple serial (RS-232) links which are available on most vision systems. This solution requires considerable extra effort and cost, because software for the micro must also be developed. If a suitable cell control computer already exists, this system might take on the role of quality data communications with the IQS.

As regards direct process control, many vision systems are fitted with digital outputs, which can be programmed to communicate directly with process control electronics, or drive relays to operate sorting equipment on the basis of inspection results.

11.7 ULTRASONIC TECHNIQUES

The area of non-contact inspection is dominated by machine vision techniques, but there are a number of other non-contact methods for obtaining dimensional and surface contour information. Eddy current distance sensing for conductive materials, and the use of laser ranging probes with CMMs are outlined in other chapters. Ultrasonics techniques are a possible alternative to machine vision for some applications, although their scope for precision measurement is limited.

Ultrasonic sensing has been the subject of considerable R&D spending in recent years, and a number of QC applications have been developed, particularly in the area of non-destructive testing (NDT). Ultrasound may be used for both imaging and ranging. Technology in both these areas is highly advanced for defence (sonar) and medical (scanning) applications, but the resolution required from dimensional inspection equipment is frequently impossible to achieve using ultrasonic techniques.

Another problem is the medium of transmission. Ultrasound travels well in solids and fluids, and through body tissue, but sensing in air (where we usually wish to inspect items) presents major problems. However, transducers have been developed which allow practical systems to be used for the high-speed sorting of small components, according to the ultrasound 'signature' transmitted by or reflected from the objects.

An alternative to ultrasound transmission through air is the use of surface acoustic waves. These have been used to examine the condition of the edge of presswork components. The main quality problem with presswork (apart from occasional punch breakages) is the formation of burrs on edges, as the tooling wears. To detect burrs, ultrasound is transmitted on to the surface of the part, and the reflection of this sound from the sharp edge of the part gives a clear indication of the extent of wear.

11.8 CONCLUSIONS

Machine vision technology is no longer confined to the research laboratory, and is playing an increasing role in manufacturing. There are a multitude of applications in inspection and process control, and vision systems are natural data acquisition elements for an IQS.

Machine vision requires an ordered inspection task and environment, particularly with respect to the presentation and illumination of the products to be inspected. Most applications require special lighting. Inspection speed for real-time operation with high-speed processes or complex products may be a problem, because the processing of an image requires a large number of time-consuming computations. For this reason most industrial machine vision systems incorporate powerful high-speed computer hardware, and are consequently much more expensive than, for example, general-purpose microcomputers.

The development of practical inspection applications requires a considerable investment because of the complexity of the engineering and programming needed. Demonstrating the cost effectiveness of the proposed solution will often be a problem. It is usually much more sensible to adapt an existing solution to a straightforward problem than to attempt a state-of-the-art development in an uncharted area.

12

Automatic testing in the electronics industry

12.1 INTRODUCTION

'Automatic test equipment' (ATE) is the generic name for the inspection and test equipment used in electronics manufacture. The metrology traditions of the metal cutting industries have limited relevance in the electronics industry, although, of course, many components do require dimensional measurement. The majority of the quality features which must be inspected relate to component, sub-assembly, PCB and system electronics.

Discrete electronic components may be tested before assembly – usually to a PCB. A sequence of checks on bare and assembled PCBs can be run, as well as the functions of a completed sub-assembly or system. These testing activities are extremely time-consuming and subject to error when performed manually, because of the complexity of the task. For this reason the technology for automatic test is well advanced, facilitated by the electronics and computing skills of personnel in this industry.

12.2 AUTOMATIC COMPONENT TESTING

Although some individual components, such as resistors, may be inexpensive, the cost of identification and rectification will be high if a faulty component is assembled in place. It is notable that certain large electronics manufacturers who pride themselves on developing supplier relationships and making quality the complete responsibility of their suppliers, have retained automatic component test facilities.

Automatic equipment is available for 100% inspection of simple, standard components prior to assembly. The nominal value and tolerance band of a batch of components is programmed into the testing machine and components are handled automatically at high speed, any suspect ones being removed or marked, perhaps by bending up a lead. The components must be packaged in such a way as to allow automatic handling, often using a bandolier arrangement.

Components may also be automatically checked on assembly to the PCB. A number of automated and semi-automated board stuffing systems incorporate some degree of automatic verification of component presence and value.

Integrated circuits (ICs) demand much more complex component testing strategies. Custom devices, known as application-specific integrated circuits (ASICs) may be tested using specially built test boards and a test program, probably administered by a suitable computer. The IC designer will probably also be responsible for the design of suitable test equipment. The cost of specialized equipment for very large scale integration (VLSI) devices may mean that only the manufacturers will possess equipment able to exercise the IC completely. Yield rates for ICs are frequently poor, making 100% inspection essential.

12.3 IN-CIRCUIT PCB TESTING

ATE is widely used for in-circuit PCB testing (see Figure 12.1). This type of testing checks the assembled board in detail for faulty, missing, reversed or incorrect components, rather than for overall functionality. In-circuit testing is very effective at isolating manufacturing quality problems and providing diagnostic information for rework activities. This type of testing is frequently used for high-value PCBs, where rework of defective boards is economically justified.

In-circuit testing equipment usually uses a 'bed of nails' – an array of spring-loaded probes which contact the conductive nodes of the PCB. A test fixture consists of a matrix of gold-plated probes arranged on a back plate, in a pattern which is unique to a particular PCB. A typical fixture for a complex PCB will hold 400-500 probes. In some systems the PCB is held on to the contacting probes by the drawing of a vacuum underneath it. The connectors at the edges of the PCB, which are used to connect it to other components, can also be used to conduct tests.

Each one of the probes must be connected to the ATE computer,

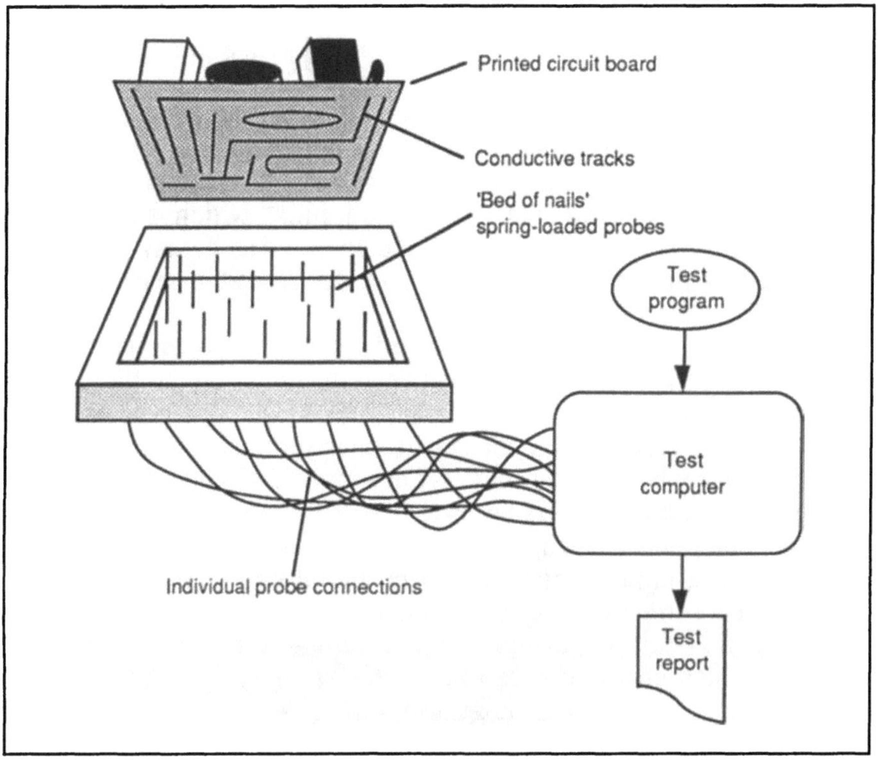

Figure 12.1 In-circuit PCB testing using ATE

which involves a mass of individual wires, and specialized I/O abilities. The design of the fixture and development of the associated software is a major task for a complex PCB. Equipment of this type is usually sophisticated and expensive, and ATE manufacturers will often undertake the design and construction of PCB contacting systems for the client manufacturing company.

The PCB design has a great influence on the testing process. Many PCBs contain connection nodes and components which are included purely for test purposes, and have no functional purpose. This obviously increases manufacturing cost, which must be offset against the appraisal and failure cost savings resulting from more complete and rapid test.

The latest SMT has increased the problems of in-circuit fixture design, because full nodal access for connection to the probes is more difficult. Fine pitch ICs with pin spacings of only 0.025 in. are becoming common, as are PCBs with eight or more layers, the inner ones only being accessible to probes if special plated through holes are included in the design as contacting nodes.

Once the PCB is attached to the fixture, a sequence of tests is performed by the ATE computer, which will require a test program with specifications for each different PCB. The tests are conducted by the application of voltages to selected pins, which should give particular results at other points of the PCB. The actual results are monitored at the appropriate pins, and the results assessed by the computer.

The sequence might start with a simple 'shorts and open' test, which checks the electrical connectivity between the nodes on the PCB, using a threshold level of resistance of perhaps 30-40 Ω. This type of check may be run on bare boards, but is particularly valuable after soldering has taken place, because it can detect dry joints and solder bridges.

The test system will next be used to test the presence and value of analogue components such as resistors and capacitors. Some analogue tests require a number of nodes to be monitored simultaneously to ensure that a correct reading is obtained.

Finally the board is powered up, and digital tests performed on IC function. Each component can be tested by applying voltages to the appropriate input pins and monitoring the output pins. Libraries of test procedures for standard ICs (based on the truth table for that component) will be available as part of the ATE software, and can be incorporated into the test program. Back-driving the outputs of other nearby components with voltages which could be harmful must be avoided.

The entire test sequence might take three minutes, compared with several hours for manual inspection of the same PCB. Occasionally, in practice, the ATE computer may indicate the presence of a spurious fault which will probably be related to poor contact between PCB and the test probes. Perhaps 5% of identified faults will be these phantom failures. The operator will attempt to clear the problem, perhaps by relocating the board on the fixture.

When genuine faults are detected, the ATE computer will provide a screen display and printout indicating the location and type of fault. This information is vital to the technicians who must debug the problem. The detail and accuracy of this diagnostic information can make a great deal of difference to the ease with which the identification of the cause, and its rectification, can be carried out. As production may be halted until the cause of a fault is determined, the influence of the integrity and quality of this information extends beyond the rework benches. The most frustrating (and far too common) experience for the debug personnel is to be unable to find a fault where one has been indicated by the ATE. This casts doubt on the accuracy and validity of the automatic test process.

Machine vision systems are also used for the inspection of PCBs. Specialist systems are available to inspect anything from artwork registration and type number on bare boards to glue spot, component type, placement and solder quality. It is usually necessary to examine both sides of the board.

Useful as they are, neither the 'bed of nails' type of testing equipment nor a vision system can determine the full functional integrity of a PCB, let alone a complete product composed of a number of boards and other components. Functional testing is required to ensure customer satisfaction.

12.4 FUNCTIONAL TESTING

Functional testing of electronic products and sub-assemblies will usually be undertaken as a final inspection. Functional testing provides a relatively low level of component diagnostic data, but can detect subtle functional difficulties caused by the interaction between components, such as timing problems. Two types of test are commonly automated.

The first is known as 'hot-box' testing. Here the PCB or other sub-assembly is attached to a slave product or system which is lacking only that component. Special connectors are often provided to ease the connection process and prevent damage and wear to the slave system. The function of the complete product is now tested to exercise the component under test as part of the system. A computer may be used to sequence and administer the tests required, and to record the results.

A typical example is the testing of a tape drive control board by connection to a slave tape drive unit, in turn connected to the computer which will run the test sequence. One or more tapes will be loaded, and a program of reading and writing data to and from the tape will verify the performance of the control board under operating conditions. If faults are detected, the computer will log them and generate a failure report to aid the debug personnel.

Hot-box testing is often used as the only test for low-value PCBs, such as those in consumer electronics, or in combination with in-circuit testing for more complex boards. In the latter case, a failure rate of 3-5% at functional test after in-circuit testing would be commonplace, and demonstrates the inadequacies of in-circuit methods.

The second method is perhaps the most common type of electronic test, and is often called 'rack and stack' testing. This involves use of a range of instruments and associated equipment such as signal generators to simulate a range of input conditions. The sub-assembly or system outputs are then assessed against the tolerance limits for that parameter. Adjustments may be performed at this stage if necessary. This type of quality testing is very widely used, but can be very time-consuming, and require highly skilled personnel, when performed manually. Automation has made considerable progress in this area.

A wide range of automated electronic test equipment is available which may be used for quality testing. Much of this equipment will be connected to a test computer via a standard interface (variously known as the IEEE-488, GP-IB or HP-IB) and remotely programmed to carry out the required tests. It is normal to find companies developing their own test and inspection benches using a suitable selection of programmable instruments linked to a PC or workstation. Software is available to assist with the communications and instrument programming task. A library of individual tests is built up, and the relevant ones selected for each product using the appropriate tolerance limits for each quality characteristic.

For example, an optoelectronic device such as a miniature data communications laser module will be connected to both optical and electrical programmable instruments. A range of perhaps 15 individual tests will be performed in quick succession under the control of a workstation which is interfaced to each instrument. The tests might include optical output power, sensitivity, threshold current, central wavelength, spectral width, and temperature sensor response. The results will be assessed against tolerances and any non-conformance indicated. A detailed quantitative report will be printed if necessary, or archived to disk.

Before or during testing a period of temperature stress (known as 'burn-in') can be used to accelerate the failure of any dubious components, such as those which might have received damage caused by static electricity during handling.

Long-term testing of equipment to identify faults which may arise in service is carried out as a further extension of the QA function by many electronics manufacturers. The products are usually placed in racks and subjected to continuous usage, sometimes with additional temperature stress. Military and aerospace electronic equipment will typically be subjected to much more thorough testing than consumer products.

12.5 AN EXAMPLE OF INTEGRATED ATE

A description of an application which emphasizes the importance of integration and effective quality data management to driving the improvement process is provided by Dowding (1987). In this case study he describes how Plessey Networks and Office Systems manufactured digital telecommunications equipment incorporating VLSI devices and custom hybrid circuits. The latest SMT was used to mount these components on to complex PCBs by automatic placement methods.

The traditional approach to manufacturing quality at Plessey had relied on a paper-based QC system, and on the ability to rework and retest PCBs at low cost. First-time test pass rates on established products were of the order of 60-70%, while for complicated new technology products increasing difficulty was being experienced in achieving even these levels. Two-thirds of the quality costs of the business were attributable to manufacturing.

A team representing engineering, production, QA, test and business systems staff was formed to specify and develop a computer-based quality data collection and management system, using a TQM approach based on process improvement.

The paperless IQS which was developed is a two-level combination of the Hewlett-Packard QDM quality management system to provide a long-term quality database, and CAPTAMS, a real-time data management system from CAP Industry Ltd.

The CAPTAMS/QDM system at Plessey incorporated four Marconi ATE units together with other test equipment from Hewlett-Packard, Factron and ATE Systems International. Manual entry of quality data to the system was possible via 35 terminals. The terminals could also be used with CAPTAMS to provide immediate on-line fault diagnosis and rework information, board status data and product tracking. Real-time warnings to the process operators alerted them to problems, and information for process control and defect prevention was available immediately.

The QDM side of the system provided graphical SPC control charts, Pareto diagrams and histograms as well as tabular reporting. This output was used for process improvement. Because of the complexity of the manufacturing operation, there were interactions between different process stages which had not previously been isolated. These could now be identified, and measures taken to improve quality performance.

One result of the newly available plant-wide quality information was the reduction (and elimination in many cases) of goods-in testing. In manufacturing, first-time pass rates were improved to 85-90%, and even new products reached pass rates of 70% within a short period. Dowding estimates that the financial benefits of the system were in excess of 2% of sales per annum.

The less easily quantifiable benefits were also considered important. The IQS was 'owned' and operated by production staff, and the shop-floor personnel became more highly motivated to achieve better quality performance.

12.6 SUMMARY

In the electronics industry, a variety of test equipment has been developed to help achieve quality objectives in the face of considerable technical problems which have meant that quality levels are often quoted in terms of yield rates, rather than defect levels. 100% test and rework of defective production is very widespread.

Testing is performed on individual components before assembly, and on bare and assembled PCBs. Functional testing on finished products and sub-assemblies is also required. Because of the large numbers of individual components, and the complexity of PCBs and finished products, ATE has been developed for all stages of the manufacturing process.

For automatic testing of simple standard components, high-speed test equipment is available. More complex devices, such as ICs, require special test boards and programs, which may be available only to the IC manufacturer. PCB testing is carried out with specialized in-circuit test equipment which uses a 'bed of nails' – a matrix of contacting pins which connect the PCB to the computer, allowing it to test the board for a large number of potential faults. For functional testing, the ATE is based on general-purpose computers to which are interfaced a variety of programmable instruments. A sequence of tests will be performed under program control. Another method of functional testing is to connect a sub-assembly to an otherwise complete slave system and run a sequence of checks on the operation of the complete system.

The ATE detects non-conformance, but also has an important role in the provision of diagnostic quality information. The equipment must be able to identify, if possible, the actual component or connection which is causing the trouble, so that rework or 'debugging' can be performed economically. This defect identification function can provide excellent information for prevention and improvement.

ATE can readily provide output of test data for all products inspected. It can also often accept test programs and specifications from a remote computer. These are the major requirements for local quality data collection units in an IQS, and thus ATE is very suitable for integration into plant-wide systems. Electronics manufacturers have taken the lead in this area, because of the very large amounts of test data which are collected by ATE. Integration efforts have been facilitated by the electronics and computing skills which are available in these companies.

13 Automatic dimensional gauging

13.1 INTRODUCTION

Many different sectors of industry have developed automatic systems to detect non-conformance and trends, and exercise process control over discrete part manufacture. Dimensional inspection is perhaps the most important area, and here there are established methods using dedicated hardware for the inspection of machined, moulded and fabricated parts. The technology consists of contact measuring equipment specifically designed to operate on the shop floor, and often – perhaps rather misleadingly – known as 'dimensional gauging'.

In addition to dimensional gauging, there are significant associated areas, which will not be mentioned further here. These include the assessment of surface finish, roundness measurement and hardness testing. Automated equipment is widely available for all these purposes, and can be linked to an IQS.

13.2 SHOP-FLOOR DIMENSIONAL INSPECTION EQUIPMENT

Simple electronic comparators have long been a common sight on the inspector's bench. The typical instrument consists of a single measuring transducer, usually fitted to a stand incorporating a small surface plate, with the electronics and digital or analogue (dial) display incorporated in a separate box. This type of equipment (as a comparator) must be set up to a reference standard, usually an accurate gauge block of the nominal dimension, and can then be used to inspect similar dimensions on components.

Inspection instruments of this kind involving some level of

automation have been available for many years. The lowest level of automation is direct indication of conformance to preset tolerance limits. This is often achieved using a sequence of three indicator lights, showing below-tolerance, within-tolerance and above-tolerance conditions, or different bands for selective assembly purposes. The use of the term 'gauging' originates from this type of equipment, because like the traditional metal go/no-go gauge the instrument gives conformance information rather than a numerical result. This degree of automation was common before the advent of digital electronics. It is relatively easy to link such equipment to automatic sorting devices.

Multiple dimensions may be simultaneously inspected, using a number of compact transducers. The transducers will be mounted in a multiprobe inspection fixture which also locates the component, so that they contact the component in the correct places. This usually limits the use of each inspection fixture to a single component, although most systems allow a certain amount of adjustment for the transducers in the fixture. More complex fixtures may have a variety of special mechanisms, such as sliding elements, to enable the probes to contact the chosen measuring points.

Multiprobe systems are initially calibrated with a master setting workpiece which has accurately known dimensions. Instruments of this type contain no absolute measuring scales, and operate as comparators. The calibrated master is placed in the measuring fixture and each probe adjusted to the centre of its measuring range, hence zeroing the output. This calibration process must be repeated at regular intervals. Naturally the master must also be calibrated with a CMM or traditional metrology instrument, to maintain traceability to national and international standards.

Figure 13.1 shows an inspection fixture for a complex component of a gas turbine aero-engine. The fixture consists of a location for the component and an arrangement of compact linear variable differential transformer (LVDT) dimensional transducers. The transducers are brought into contact with the part, and indicate the dimensions of the gauging points with reference to the master setting component.

In common with most multiprobe fixtures, this one requires moving elements which open up to allow the component to be loaded and unloaded. Here the transducers are mounted on three slides, which are moved up to fixed stops by pneumatic actuators when the part has been clamped into place. Other actuators then move further transducers into their measuring positions. Both sliding elements and the alternative methods of closing inspection fixtures, such as pivoting hinged elements, can be prone to wear or misalignment.

Figure 13.1 A multiprobe inspection fixture

13.3 DIMENSIONAL MEASURING TRANSDUCERS

Another problem is the limited measuring range of those types of transducer which have the necessary resolution for precision measurement. Dimensional transducers may be of contact or non-contact varieties. Contact transducers are more commonly used, and are often called 'probes'. These are usually of the LVDT type although half-bridge and capacitative probes are also used.

The LVDT consists of a primary wire coil wound over the length of the probe. An alternating current is fed through the primary coil from a separate source. The frequency of this supply will be anything between 50 Hz and 20 kHz. Quartz locked circuits are usually used to ensure a constant frequency, which is important for accuracy. Two further secondary coils are arranged in the transducer, such that an output voltage is induced if a magnetic core placed between the coils is moved from a central position. This output voltage is then amplified and the results displayed in terms of the movement of the core, which is connected to the probe tip.

Linear measuring ranges vary from 7.5 mm down to 0.5 mm. With measuring spans like this, probes must usually be adjusted to the centre of their range just in order to cover the tolerance band of one component, let alone several similar parts. Different LVDT sizes and configurations, including a side acting variety, and several types of contacting tip are illustrated in Figure 13.2. Measuring forces are of the order of 60-100 grams at the central zero position.

LVDT resolution is theoretically infinite, with linearity the limiting factor. Typically this is 0.1% or 0.2% of the range (repeatability is of the same order). The difficulties which a small measuring range causes are illustrated by some work carried out by the author in far-off pre-digital electronics days. The project involved the use of an LVDT transducer in order automatically to inspect the profile of the eccentric lobes on diesel engine camshafts, where complex and severe quality problems were being encountered. This task had taken several hours to perform using manual equipment, and hence only a small sample of the camshafts could be inspected, making it difficult to collect enough quality information to identify the causes of the problem.

The limited range of the LVDT probe made it impossible simply to turn the shaft while recording the dimensional signal from the transducer, a technique often useful to check concentricity of diameters. This was because the 'lift' of the cam lobe was far greater than the relatively tiny errors in form and angular disposition which had to be measured. Here the cam lift was more than 13 mm, while the tolerance on lift was ±0.05 mm. The solution was to build a measuring machine incorporating a master cam profile which displaced the whole LVDT automatically while it was measuring each cam lobe. Only the differences between the two cams were recorded on the pen-recorder.

An alternative to the LVDT transducer, which has recently been miniaturized sufficiently for some purposes, is the Moire fringe photoelectric linear encoder. Fairly compact absolute measuring units with ranges of up to 100 mm, combined with good resolution (typically ±0.0005 mm) are available. This type of digital transducer eliminates problems of analogue electronics and A/D conversion by incorporating a digital interface in the transducer. Transducers of this type (also used on hand-held instruments) are still relatively bulky and

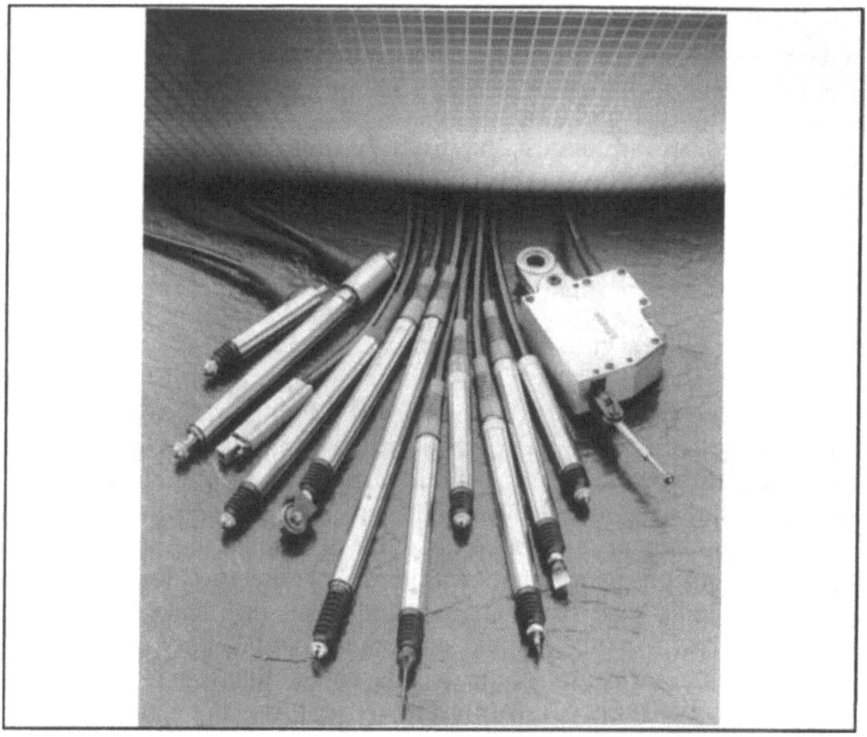

Figure 13.2 LVDT transducers for dimensional gauging *(reproduced with permission of Thomas Mercer Ltd.)*

expensive, making them impractical for many multiple gauging applications. However, the trend towards miniaturization may eventually mean the end of the LVDT and similar transducers.

Systems using these digital transducers are usually supplied with a separate control box, incorporating digital readout and a few useful features such as preset tolerance limits. For more complicated data analysis the control box often incorporates a computer interface. Unfortunately, this is frequently a binary coded decimal (BCD) rather than the more convenient serial (RS-232) or parallel (IEEE-488) interfaces actually found on micros. To design, build and test a custom interface to convert the BCD output to one of these more useful standards can take a competent technician weeks, as the author has discovered to his cost.

One important type of non-contact transducer operates on the eddy current principle, and can only sense a conductive component. The maximum sensing range of these transducers varies from 0.5 mm to 80 mm, with a typical accuracy of 0.2% over the central, most linear, part of the range; this is equivalent to that of an LVDT. Temperature stability is excellent, as is the ability to measure transient motion such as vibration. Complete range, displacement and thickness measuring systems are available for single- or multichannel operation, with the control units capable of simple analysis and recording, and provided with a computer interface. Despite their many applications in inspection, eddy current methods are more often used for position sensing and control of machine elements than for product sensing.

Another non-contact method uses an optical range transducer which incorporates a tiny laser diode to send a beam of light angled towards the surface to be sensed, which is reflected and focused onto a photosensitive detector. One such product is the Renishaw system which can be used in place of a touch trigger probe on a CMM, and also on other non-contacting inspection systems. A resolution of 1 μm is claimed, with a realistic accuracy of ±10 μm, over a measuring range of 4 mm. The ability of the system to take 50 readings per second 'on the fly' allows a surface to be scanned very rapidly with a CNC CMM.

13.4 AUTOMATED APPLICATIONS USING DIMENSIONAL GAUGING

A wide range of computerized shop-floor gauging systems, including hardware and software, is available from metrology equipment suppliers. Some systems are based on stand-alone micros, while others use a distributed system with central data storage. The use of a computer system allows quality data to be collected automatically and compared with the nominal and tolerance values for each dimension of every component, which are held on disk. The measurement data may be stored likewise, or printed as necessary. SPC techniques are usually provided to allow the detection of process trends. It should be noted that some commercial systems have suffered repeatability problems resulting from poor analogue voltage stability.

Much existing electronic gauging equipment may be used with a computer by converting the analogue output from the LVDT transducers to digital form. This is easily carried out using an analogue to digital (A/D) converter, a fairly inexpensive piece of electronic equipment which is available to suit virtually all types of microcomputer. Many older items of electronic inspection equipment have been converted in this way. When interfacing to the computer has been achieved, a program is needed to compare the digitized output from the transducers with the nominal values and tolerance limits for each dimension. The calibration process may also be largely automated if details of each master are stored on disk.

Automation and integration of inspection activities using shop-floor gauging is fairly straightforward compared with other methods, such as integration of a CNC CMM. Full automation would be a serious option in a dedicated manufacturing cell, where product changes were rare. This is the actual situation in some so-called flexible manufacturing systems, where expensive CMMs may be found inspecting a small handful of different components. This degree of flexibility could often be handled using a few inspection fixtures interfaced to a single micro, with significant savings.

Apart from the relevant inspection fixture(s) and interfaces, the vital component of such a system is the computer, running suitable software. This could be a cell control computer, or more probably a micro dedicated to the inspection task. The latter could be integrated into an IQS, receiving specifications and reporting results to a central database, in the manner described in Part Two.

In the manufacturing cell, full automation often requires a robot work-handling device. It is usually fairly easy to interface an industrial robot to a micro. The author once produced a fully automated demonstration system for inspecting six diameters on turned parts in less than one week. The system used a multiprobe inspection fixture interfaced to an elderly micro, which was linked by four digital I/O lines to a Puma industrial robot. The micro signalled the robot to insert components into the inspection fixture, captured the data from each probe and then assessed conformance to specifications. The robot was then instructed to place the component (according to the results of the assessment) in a good or reject parts hopper. Bar graphs and statistics were provided by the micro, and could be printed as desired.

A practical example of a fully integrated system along the same lines can be found at a Rolls-Royce aero-engine plant. Here a fully automated broaching line, served by a number of robots, is engaged in the machining of the precision root form of gas turbine blades. The inspection station is the final station in the manufacturing sequence. A robot loads every component to a multiprobe inspection fixture. The dedicated computer system examines the results from every measured dimension, comparing them with tolerance values. The disposition of each component by the robot depends on the inspection results. A link to the cell control computer allows work to be halted if necessary. Simple statistical analysis is also performed by the system.

Many applications do not require automated work handling, but otherwise offer all the advantages of automation. Further examples of computerised systems at Rolls-Royce may be found in the inspection of nozzle guide vanes at the casting stage, and for finished low-pressure vanes. Both these parts are in contact with the gas flow through a jet engine, and have complex curved aerofoil sections.

In the foundry system, each type of investment cast component has a special inspection fixture, incorporating up to 25 transducers. These fixtures are substantial and expensive items; each is mounted on its own trolley, which must be wheeled up and connected to the computer when required. The full inventory of inspection fixtures required has represented a large investment, and takes up a significant amount of floor space. The computer system assesses conformance and provides the usual statistical and graphical displays.

The low-pressure vanes were inspected using a dedicated multi-probe fixture connected to a computer for some years. More recently a CMM has been introduced as an alternative inspection method. The CMM is slower, but has more recent and comprehensive software, valuable for analysis of the results obtained from the complex component geometry. Despite occasional trouble with sticking probes, however, the older system is still in use and provides a rapid method for 100% inspection and acceptance testing.

13.5 PROCESS CONTROL INTEGRATION

One important type of integration which is available in some systems is feedback to control the manufacturing process. As mentioned previously, CNC machine tools have tool offset tables held in memory, which inform the machine's control system where its tool cutting edges are located. These tables may be altered to compensate for tool and cutter wear. If a wear trend is detected from measurements, an automated gauging computer can compute and propose (or even directly implement) changes to the tool offset tables.

One project which was supervised by the author involved the development of a prototype system of this type. A multiprobe inspection fixture was interfaced to a microcomputer, which was itself connected via a simple serial (RS-232C) link to the Siemens controller of a CNC turning centre. This controller treated tool offset tables in just the same way as its part programs, and could easily be persuaded to download the offsets to the micro.

The software which was developed for the micro had the ability to capture information from the inspection fixture, compare the data with specifications, and, when sufficient data had been collected, to generate control limits on an SPC control chart for sample means. If either a specification or control limit was breached, or a trend detected according to a simple algorithm which examined the position of the last seven points on the chart, the micro could propose amendments to the tool offset table, on the basis that wear had occurred. (If a sudden large change was detected, it was assumed that a tool breakage had occurred and no attempt was made to compensate with tool offsets.) If the operator accepted the changes, the amended table would be downloaded to the machine tool controller.

The complexity of a practical system of this type occurs because the exact features and dimensions produced by each cutting tool must be known to the software, so that it can identify which tool offset to alter.

One successful application involving large-scale use of computerized gauging with process control integration is to be found at a Bendix manufacturing plant. Here a section of CNC lathes machine automotive brake components, with typical batch sizes of several hundred.

For each batch a multiprobe inspection fixture is built up from a drawing, using a standard kit of parts. The fixtures involve a framework, locating pieces and workpiece clamps. Transducers are fitted to the framework and to pivoting elements which close around the component to allow measurement on all sides.

Once built, the fixture is connected to a micro located near each lathe, then tested and calibrated with a master component having known dimensions. When a part has been machined, it is placed in the inspection fixture by the machine operator, and the fixture closed. All dimensions are simultaneously inspected and the results are graphically displayed as bar charts on the screen, giving conformance information directly to the operator.

Mean and range SPC control charts are also immediately available, as are details of any changes required to the tool offset tables (which must be manually amended). Finally, inspection results from each batch are sent to a central micro where they are archived, and from which printed reports are available. This system has proved satisfactory over several years of hard usage, much of the time working continuous shifts.

13.6 CONCLUSIONS

Shop-floor dimensional gauging equipment is a well-established technology, with thousands of applications in many areas of industry. It involves the use of dedicated inspection fixtures fitted with a variety of contacting and non-contacting transducers to measure several dimensions of a component rapidly. These transducers can be interfaced to a computer; and by analysing and processing the quality data obtained, the benefits of automation and integration can be realized just as effectively as with the use of more recent and fashionable techniques.

The main rivalry to such specialized inspection equipment comes from CMMs. For small quantities, frequent product changes and complex parts CMMs tend to be superior. For large quantities, special hardware in the form of inspection fixtures is potentially more economical, and can save inspection time, while allowing all the benefits of computer automation and quality data integration. Automated dimensional gauging equipment is suitable for use as either computer-aided or fully automated inspection stations, depending on the computer communications facilities available and the work handling arrangements.

However, complex special-purpose multiprobe inspection fixtures are expensive; those containing several tens of probes are comparable in price to a CMM, and they are capable of inspecting only one component. The manufacturing lead time will be many months, while a CMM can be obtained virtually 'off the shelf'.

Part Four

Quality data analysis and management

In this final part, techniques of data analysis and management for the integrated quality system (IQS) are the topic. The concept of the IQS, described in Chapter 8, includes a strategy for data collection from manufacturing, using local quality monitoring computers and computer terminals. In order to provide plant-wide information on quality performance these are linked to a central quality database which can collate, store and analyse – in other words manage – the quality data, to provide decision support and routine quality reporting functions.

The technologies for automated quality data collection which were described in Part Three will provide much of the quality data from manufacturing required by the database. Direct entry of inspection data via terminals can supply much of the remainder, and also provide other essential quality information, including items such as supplier performance, instrument calibration, non-conformance control and corrective action tracking.

There are a number of purpose-designed quality database systems available, and it is also possible to develop quality database applications using general-purpose database software.

Before going on to describe the quality database, however, we must examine two current trends in the automation of quality data management. The first of these is well established, and has been referred to on many occasions already: automated statistical process control (SPC). SPC functions are incorporated into the software of many specialized quality data collection systems, particularly coordinate measuring machines (CMMs). Automated SPC is also available in the form of software to run on general-purpose computers, and can assist both process control and the provision of information for improvement. Some SPC software products also incorporate a database with quality data management functions.

There is also a tendency for direct entry shop-floor data collection

(SFDC) systems to include collection and collation of quality data, in addition to the usual functions of material tracking, machine utilization and operator monitoring. At present, the quality data management abilities of SFDC systems tend to be rudimentary, but it is possible that they may, in time, develop to fulfil many of the requirements of the quality database.

A further theme of this part is computer-aided inspection (CAI), an element of the IQS which incorporates quality data collection and analysis functions. Many CMMs, automatic test equipment and dimensional gauging systems currently operate as CAI stations, and the more advanced automated SPC software is also moving towards convergence with the CAI idea. A case study is provided of a low-cost, general-purpose CAI workstation for interactive dimensional inspection in a batch production environment. The system described uses a variety of digital hand-held measuring instruments to acquire quality data directly, and provides process control information using SPC techniques, as well as batch summary information to a central database.

14

Statistical process control software, data collection and computer-aided inspection

14.1 INTRODUCTION

This chapter describes the application of automated SPC, SFDC and CAI. Only a small proportion of inspection operations lend themselves to full automation in quality monitoring, because essential manual inspection operations cannot always be eliminated economically. In these areas the use of a computer to semi-automate the quality data collection operations is an alternative which can reduce inspection errors, provide immediate processing of process control information and eliminate inspection report paperwork.

At present, there are two trends which may be distinguished in this area. First, and most common, is the SPC software package, usually running on a micro. Second, computer terminals or portable computers for SFDC are used, linked to a database system running on a mini or micro.

Combining elements of these two themes, a new concept is emerging. This is the general-purpose CAI workstation, a flexible and relatively inexpensive system for use in those areas where fully automated inspection is not the most appropriate strategy. This chapter will review the existing techniques, and examine a case study in the development of a CAI workstation.

14.2 STATISTICAL PROCESS CONTROL SOFTWARE

As noted in Chapter 1, the principles of SPC were first expressed by William Shewhart as long ago as 1931. For many years they were little used, but now enjoy great popularity as the major group of techniques supporting a TQ approach. They are applicable throughout the business environment, in commercial and administrative areas as well as in manufacturing. It is not the purpose of this chapter to review SPC concepts and practice. Many excellent practical books have been written on the subject, such as that by Owen (1989). However, the reader should be aware that SPC techniques are based around analyses of the variation inherent in most activities.

Variation in product quality characteristics arises from influences on the manufacturing process due to factors connected with material and process conditions. These influences may be random or systematic. The latter are known as 'special' or 'assignable' causes. They originate from a mechanism which may be identified, and may have a practical remedy. In fact much variation which is considered to be random will be found to have a distinguishable cause and cure if sufficient quality data can be collected and analysed, so there is no absolute dividing line between the two types of variation.

SPC tackles variation using two main techniques:

1. Capability studies use statistical analysis techniques to compare the quality performance of a process or function with the target or specification for that function. Information for the improvement of performance and stability through the elimination of the causes of variation is the main result of these studies.
2. A type of graph, known as a control chart, is used to monitor the quality performance of a process or function. Critical quality characteristics are chosen and plotted against time. Statistical methods are used to assess the extent of the process variability being displayed in order to predict and examine trends, and thus prevent the occurrence of non-conformance by process adjustment.

When correctly used in conjunction, these two SPC techniques are among the most powerful weapons in the TQM arsenal. Certain enthusiasts, however, have a tendency to lose perspective and view SPC as a complete TQ approach in its own right.

SPC methods are usually based on information obtained from

sampling. Statistical theory allows estimates to be made of the quality characteristics of a complete batch from the inspection of random samples from the batch. Sampling inevitably introduces possibilities of error in these estimates, which can only be eliminated by 100% inspection. In addition, the use of sampling requires special correction factors to be applied when, for example, setting up control charts. Sampling, as suggested in Chapter 8, has only one major advantage: it reduces the inspection workload. When using computers for SPC, this means that a number of processes may share a single microcomputer or terminal.

It is not necessary to have a comprehensive grasp of the statistics involved in order to use SPC techniques. Capability studies and control charting techniques have been developed to minimize the calculation involved in obtaining meaningful results. Most techniques were established before the advent of the electronic calculator, and hence one of the most common types of control chart employs sample range, rather than standard deviation (which is a better measure of variation) because it is more easily calculated manually.

The introduction of effective SPC into an organization on a large scale is a major undertaking. Most people (at all levels) have no idea of the basic statistical principles involved, and no great desire to learn them. Some are afraid that their lack of mathematical knowledge and ability will be exposed if they are expected to use SPC techniques. A comprehensive training programme is essential to counter this problem.

The most effective SPC approaches originate at the highest level in the organization. Many companies which have had the most successful experience of SPC have ensured that full-time training for all personnel, starting with the chief executive, is mandatory. The existence of committed SPC 'champions' to drive the introduction of SPC is another vital ingredient. Without training and drive, SPC is unlikely to be a success.

A variety of computer software to perform all SPC functions has been available for some years, but some experienced SPC trainers insist that SPC methods should first be learnt and applied as manual techniques, because only in this way can the basic principles involved be appreciated. It has also been suggested that a capability study or control chart produced by a computer generates less of a sense of 'ownership' of the information displayed than a paper graph on which one's own grubby thumbprint is evident!

Despite these reservations, there can be no doubt that SPC software has proved useful to many companies, and is a profitable business for software houses. It is here to stay, and for many people a software package will be the first experience of SPC techniques. What, then, should we expect from SPC software?

The typical microcomputer SPC package will implement a full range of control chart techniques for attribute and variable quality data, see Figure 14.1. Apart from standard Shewhart charts, there are a variety of other types of control chart which may be useful, particularly the cumulative sum (cusum) chart which is more effective in showing process trends. There are a number of more esoteric types of chart, however, which have a very limited application and do not add much value to a software product.

The control charting techniques are applied on-line to the manufacturing quality data as they are collected by sampling from the manufacturing process. A number of processes will probably have to share a single computer or terminal, which means that the software must be able to store the quality data and relevant charts for one process, and switch to the data and charts for another as the time comes for new sample data to be entered. Each user will enter data and review the appropriate chart to find out if process adjustment has become necessary. The software should also provide screen displays and printouts to show batch statistics such as mean and standard deviation for the charted quality characteristics. Because these are based on sampling they are liable to error, and cannot indicate the number of defective items which have been produced, if these were not all included in the samples taken.

The other vital function of SPC software is a comprehensive ability in process capability studies and descriptive statistics, to be used off-line as a guide to process improvement, rather than for immediate process control. Graphical displays showing histograms of statistical distributions are essential. A large number of statistical tests may be performed on data, including mean, median, range, standard deviation, normality, kurtosis, skewness and various coefficients of process or machine capability, as shown in Figure 14.2. The expert user, when investigating quality problems, will find that the more tests are available the better.

Unfortunately, not all software provides accurate statistics. Some lower-cost packages tested by the author introduced significant numerical errors. The Ford Motor Company, one of the prime movers of SPC, has produced a quality standard for SPC software which is considered acceptable for use by their component suppliers.

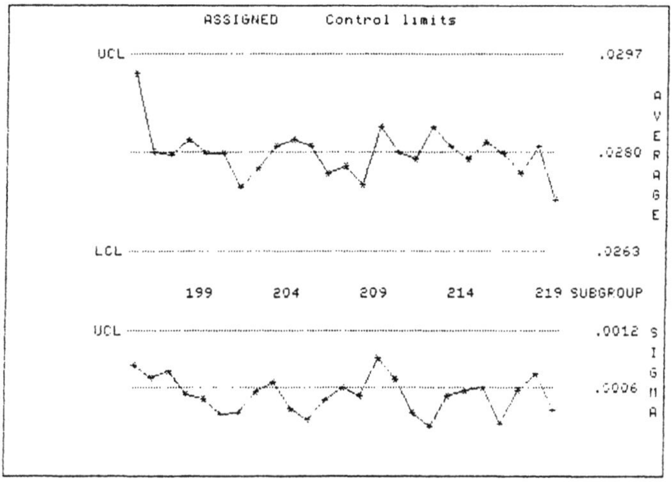

DATE	TIME	TOOL NO	MACH	UN	LOT NO	REEL #	AVERAGE	RANGE/STD	CND
71586	12:40 AM	832	C1	01	0118728	141	.0294	.0008	0
71585	2:10 AM	832	C1	01	0118728	146	.0280	.0007	0
71586	3:06 AM	832	C1	01	0118728	151	.0280	.0008	0
71586	4:10 AM	832	C1	01	0118728	156	.0282	.0005	0
71586	5:47 AM	832	C1	01	0118728	161	.0290	.0005	0
71586	7:00 AM	832	C1	01	0118728	166	.0280	.0003	0
71586	2:11 PM	832	C1	01	0118728	171	.0274	.0003	0
71586	5:20 PM	832	C1	01	0118728	176	.0277	.0006	0
71586	6:21 PM	832	C1	01	0118728	181	.0281	.0007	0
71586	8:09 PM	832	C1	01	0118728	186	.0282	.0004	0
71586	9:36 PM	832	C1	01	0118728	191	.0281	.0003	0
71686	1:13 AM	832	C1	01	0118728	196	.0276	.0005	0
71686	3:28 AM	832	C1	01	0118728	201	.0278	.0006	0
71686	4:48 PM	832	C1	01	0118728	206	.0275	.0005	0
71686	5:51 PM	832	C1	01	0118728	211	.0284	.0009	0
71686	10:06 PM	832	C1	01	0118728	216	.0280	.0007	0
71786	12:02 AM	832	C1	01	0118728	221	.0279	.0003	0
71786	1:01 AM	832	C1	01	0118728	226	.0284	.0002	0
71786	2:08 AM	832	C1	01	0118728	231	.0281	.0005	0
71786	4:04 AM	832	C1	01	0118728	236	.0279	.0006	0
71786	5:25 AM	832	C1	01	0118728	241	.0282	.0006	0
71786	6:43 AM	832	C1	01	0118728	246	.0280	.0002	0
71786	6:00 PM	832	C1	01	0118728	251	.0277	.0006	0
71786	7:47 PM	832	C1	01	0118728	256	.0281	.0008	0
71786	8:47 PM	832	C1	01	0118728	261	.0272	.0004	0

Figure 14.1 Control chart display from SPC software

```
     Inspection Data Analysis                    10:28 AM  MON., 29  JUNE, 1987
*******************************************************************************
PART NUMBER : 43812001                      PART DESC     : SERPENT
OPERATION   : 01 - STAMPING                 CUSTOMER      : IBM
FEATURE     : 01 - CONTACT GAP              SPECIFICATION: .0240   .0320     IN
SAMPLE RATE :      5 1/REEL                 DATE WINDOW   :  7/15/96 TO  7/20/86
*******************************************************************************
SORT PARAMETERS: TOOL=            MACHINE=        UNIT=   LOT=          OPER=
```

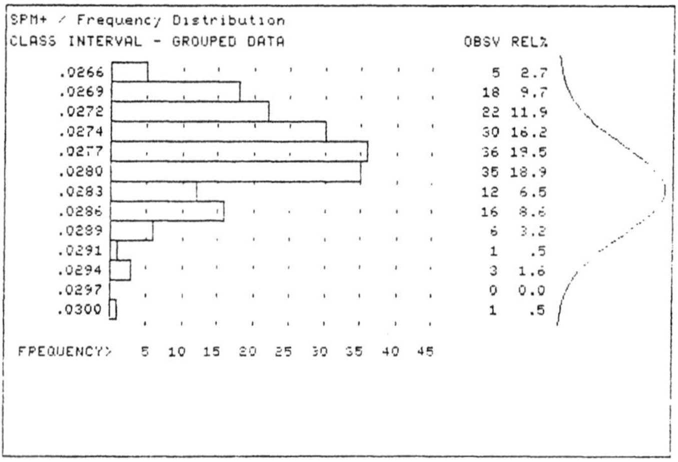

Numerical Analysis

ESTIMATES OF THE MEAN	OBSERV : 185 MEAN : .0279 STD ERROR: .0000	95% CONFD INTERVAL LOWER : .0278 UPPER : .0280	TARGET HO: .0280 t VALUE : 2.0632 MEAN<>HO : .0270
ESTIMATES OF THE VARIANCE	SIGMA (S): .0006 VARIANCE : .3889E-06 COEF VAR : .0223	95% CONFD INTERVAL LOWER : .0006 UPPER : .0008	3 SIGMA : .0019 MEAN- 3S : .0260 MEAN+ 3S : .0298
TEST FOR NORMALITY	NON-NORMAL DISTRIBUTION	SKEWNESS : .4164 SK (95%) : .3602	KURTOSIS : .3250 KU (95%) : .7204
6 SIGMA CAPABILITY STUDY	CP : 2.1381 CPK : 2.0875 % OF SPEC: 46.7704 PTS/MIL : 0.	SPEC LO : .0240 Z LOWER : 6.2626 % BELOW : 0.0000	SPEC HI : .0320 Z UPPER : 6.5660 % ABOVE : 0.0000

Figure 14.2 Capability study display from SPC software

Implementations of other quality management techniques may be available in SPC packages. The presentation of Pareto charts and Ishikawa or cause-and-effect diagrams to assist the identification and solving of problems are two useful features. Some products also include project management tools of various types to enable improvement or prevention action plans to be presented and progressed efficiently.

Some SPC packages include a full quality data management (rather than merely storage) ability. A database for product quality information, with report generation facilities, and perhaps a spreadsheet facility, are necessary for this function. An interface to a widely used spreadsheet application is also valuable. Some products offer abilities in inspection plan generation and presentation, to help the user work through complex inspection sequences.

Most SPC software obtains quality data primarily through the keyboard. A large number of sensors are now available which are suitable for interfacing to micros, and many packages also have some ability to acquire data in this way. Naturally, those systems marketed by the measuring equipment manufacturers are the most effective in direct data capture.

Microcomputer-based SPC packages vary widely, of course, in price. This partly reflects the functionality provided, but important also is the level of presentation of graphics on screen and to a printer, and the ease of use and user-friendliness of the software. Most user interfaces are designed for people without computer knowledge, but in this respect the success of the design efforts is very variable. The more comprehensive products can tend to be more confusing for the occasional user, or for those with limited knowledge of computing.

14.3 SHOP-FLOOR DATA COLLECTION

The original, and most flexible, SFDC equipment is a clipboard well stocked with the appropriate standard forms and plain sheets of paper. If the information so recorded must be entered into a computer a number of input options are possible, varying from manual input at a terminal, through automatic scanning of checklists, to OCR. Whichever method is used, this extra operation adds time, cost and the possibility of error to the data collection process.

The most popular alternative is the introduction of data collection terminals situated throughout manufacturing areas; these may be used for input of various types of useful information to a central database. Most current SFDC systems are designed mainly for collecting data on machine usage and operator activities, and for the tracking of production. Quality-related information may be collected, but usually only in the form of data indicating the numbers of products conforming and non-conforming after each operation.

SFDC systems which are designed to collect quality information often make use of small portable computers to record the information required. The small computer can later download the recorded data to a larger system for collation and analysis. This approach has been suggested by the mobile role of the patrol inspector, common in many factories. A number of such systems are available, some based around purpose-built hardware, others relying on general-purpose hand-portable computers.

The more capable portable computers incorporate SPC functions for process control information, while some can acquire information direct from measuring equipment. However, processing speed, data storage ability, poor graphical displays and limited software availability can impose considerable limitations on flexibility. In general these systems are not capable of providing sophisticated inspection plans, although they may store quality specifications.

Vulnerability to accidental damage and the poor reliability of some data collection computers have also been mentioned as disadvantages of this approach.

14.4 A CASE STUDY IN COMPUTER-AIDED INSPECTION

CAI was introduced in Chapter 8 as one of the conceptual methods for quality data collection from manufacturing. Many systems corresponding to the CAI model are in existence; these use the technologies described Part Three, including CMMs, ATE and dimensional gauging equipment. A CAI approach is also used in automated laboratory data acquisition systems.

To develop the basic CAI concept into a practical reality for use on the shop floor, the author has developed CAI system prototypes which were able to be integrated into an IQS. The general-purpose CAI workstation which resulted is a flexible semi-automated system, designed to improve inspection efficiency where fully automated inspection is either not practical or is not the most economical approach. It incorporates aspects of both SPC software and SFDC, combining the collection and analysis of quality information with reporting to a central database.

The CAI workstation consists of a microcomputer, to which are linked measuring devices for dimensional or other inspection inputs. Dimensional inspection equipment includes hand-held callipers, micrometers and depth or height gauges, as well as manually operated CMMs. There is now a wide variety of dimensional measuring instruments available with output which may be used by a computer. Other sensor possibilities include temperature, ultrasonic and force measuring transducers, chemical and electrical sensors, and the wide range of programmable data acquisition systems for electronic component and product testing.

A detailed case study of a CAI system development is now described. Before a system development was carried out, a survey of the market in SPC packages and complete quality information systems was made, and a number of promising products were reviewed in detail. These included a comprehensive system from a major manufacturer of computer systems, and some networked or stand-alone microcomputer-based systems from smaller software houses. None of the systems was able to deliver the full functionality required, especially with regard to integration with the IQS. As nothing really suitable was available it was necessary to develop a prototype system to investigate and clarify requirements for an inexpensive CAI application which would be capable of integration with a complete quality data management system.

Four fundamental aspects of the prototype design had to be considered: first, the functions required; second, the data structures (some required to be common to other parts of the integrated system); third, the user interface (bearing in mind that the potential users were not expert computer users, and were in some cases worried about tackling a computer system); and finally, the integration with, and communications to, the CQDB.

14.4.1 *Functions required*

The fundamental requirement of the CAI workstation was the ability to inspect multiple specified quality features of numbers of separate samples from batches of different products arriving at random times. This flexibility is fundamental to the acceptance of practical CAI, which must be able to handle the complexity of real manufacturing scenarios, where things (very occasionally) do not run entirely according to schedule!

The ability to acquire data not only from interfaced measuring equipment but also from the keyboard is clearly important. Many inspection sequences have elements which it is uneconomical or impossible to automate. The data acquired from the keyboard may be of variable or attribute form, and will often be an 'acceptable' or 'unacceptable' response to a catch-all inspection task such as 'Examine component for any other defects'. Measuring equipment may also break down or be in need of recalibration. The user can then enter the required data via the keyboard.

As an example of a practical inspection sequence, Table 14.1 shows the final inspection sequence of quality characteristics for an electrical connector product after assembly. The way in which the data were acquired by the CAI workstation is shown against each feature. It is important that the software allows long inspection sequences to be suspended or aborted at any stage, in case 'something comes up', or a serious quality problem is found which means the rest of the procedure becomes redundant.

Table 14.1 CAI inspection sequence for assembled connectors

Feature	CAI data acquisition method
Pin 'A' length after assembly	Direct from calliper
Pin 'B' length after assembly	Direct from calliper
Pin overall length	Direct from calliper
Twist of pins	Gauge with attribute data entry
Pin pitch distance	Gauge with attribute data entry
Straightness of pins	Visual with attribute data entry
End position of pins	Gauge with attribute data entry
Retention force of pins	Direct from force transducer
Body bend	Visual with attribute data entry
Final view	Visual with attribute data entry

Another important function is the ability to generate and display SPC control charts for data obtained from critical variable or attribute characteristics, helping the operator exercise control over the manufacturing process at local level. The most widely used SPC charts require 'control limits' to be set (as indicators of process trends and changes) using initial samples from production or from previous batch data. The latter is especially important where batches are too small to provide initial data for setting control limits based on process performance. It has been shown that SPC can be exercised on the basis of the process variability rather than purely product variability, and this opens the door for the use of SPC in the small batch production environment. (It is also possible to use chart control limits based on drawing tolerances, either directly or modified for the effect of sample size.)

The provision of detailed inspection instructions was also considered to be an essential function of a CAI workstation. In most manual inspection, drawings and diagrams are constantly used for reference, as a source of both tolerance information and of that general understanding of the product necessary for a logical approach to the inspection task. Drawings and diagrams may be supplemented by forms containing lists of features to be examined.

Graphical information can be displayed on the computer screen, to replace part drawings on paper. An existing system had been examined which used an Apple microcomputer programmed to display and print on demand a combined inspection specification and results sheet with a simple schematic drawing for each part to be inspected. The integrated drawing and word processing capability of this computer's software was used for this system, which did not have any automated data collection function.

A search for software to perform the same function on an IBM PC standard computer showed that little was available at that time. The most promising software package, purchased for evaluation, showed itself to be incapable of the required functions; at that time these were considered to be the creation and maintenance of a database of inspection plans, comprising specifications, sequences and instructions, with screen graphics associated with each part record to provide an inspection guide. The failure to obtain suitable PC software was one factor tending to encourage the adoption of a more sophisticated central database system to hold inspection plans. With recent advances in integrated text and graphics software the initial requirement could now have been satisfied.

The final vital function was integration with a central database, providing vertical integration for an IQS. When the CAI workstation was informed that a batch which was being inspected was finished, the status of that batch was changed, a summary report to the central database was dispatched, and the option of a local printed quality report offered to the user.

The system adopted for network communications with the central database allowed a large number of CAI workstations to access the computer on which the CQDB was situated. It was possible to dispatch summary quality data, and to obtain current inspection plans from the central database. The central database was automatically accessed using the active procedures described later, to accept batch summary files and provide an appropriate selection of plans for each CAI workstation from the database tables.

14.4.2 *Information design*

The design of the common data structures was tackled as part of the specification and design of a complete IQS. The data involved consisted of the following:

1. Files containing valid works order numbers with associated part numbers derived from the production schedule for the current period. This information was provided from the automated production scheduling activities and was transferred to the database for each scheduling period, in this case one week.
2. Files containing inspection plans. These incorporated inspection sequences for each product, giving quality specifications with associated tolerance information, inspection instructions and screen graphics to indicate the required feature. The inspection plans were selected and made available by the CQDB to the CAI workstation on the basis of the information on valid works order numbers.
3. Files containing quality summary reports derived from the inspection of a batch, giving details of non-conformance and statistics relating to each variable characteristic. These were generated by the CAI workstation as inspection of a batch proceeded, and were finalized and summarized when the system was informed that the batch was complete.

When the inspection plans for each period were downloaded from the CQDB, each inspection plan was placed in an individual file on the CAI workstation disk, using a special format to avoid unauthorized alteration with an editor or word processor. Files were distinguished by filenames derived from a unique identifier used by the CQDB. For each scheduling period a local file consisting of a look-up table of available works order and part numbers, together with the status of each batch, was also created at this time. Individual inspection specifications which were not in the current files could be requested from the CQDB by part number to ensure that flexibility to deal with unexpected situations was not prejudiced.

The IDEF0 modelling technique described in Chapter 7 was used to design the information needs and outputs from the workstation in the context of the overall quality information system. Figure 7.6 shows an IDEF0 diagram of the activities involved with associated data flows.

Other data – for example, the records, arrays and files containing the individual items of information obtained from measurement of each part inspected – would not be used outside the CAI workstation, as only the batch summary information was to be exported.

These data structures were implemented using a nested sequence of data arrays. The final multidimensional array contained all data from a batch inspection, holding all the separate measurements, of all the dimensions, of all the components, of all the samples taken from that batch. This information could be saved complete to disk and retrieved, in order to continue an interrupted batch inspection when using sampling methods, but was summarized before transfer to the CQDB, and for use locally in SPC charts or printed batch summaries.

An alternative approach to data management in the CAI workstation is to use a local database. Another prototype CAI system currently under development by the author's research team took this direction, using ORACLE, a relational database, running on a PC. This system was specified as a standard Microsoft Windows 3 application using the programming language C++, in the latest 'object-oriented' fashion. It has proved very difficult and time-consuming to integrate the database to the Windows 3 programming environment, and it is still uncertain whether the development will be successful.

14.4.3 *The user interface*

The requirements for the user interface were obtained and refined through a series of meetings with potential users, as the prototype CAI station software evolved. As many of the users were not familiar with computers, the underlying idea was the use of a static windowed environment (that is, of fixed position and size), analogous to a desk top, where the windows displayed on the screen represented 'cards' or documents; those which might need to be cross-referenced were displayed simultaneously, so far as possible. Different background colours were used for different types of card to help the user distinguish between them.

The first card used in the system was the main menu, offering the choice of the following activities:

1. Inspect sample;
2. Set up SPC charts;
3. Sample standard deviation control chart;
4. Sample means control chart;
5. Obtain inspection plans from CQDB;
6. Exit system.

On selection of the 'Inspect sample' option a works order number was requested by the system to identify the batch and part to be inspected. The entered number was checked for validity against the file of all current batches, derived from the production schedule. Following the entry of a valid works order number, the inspection record card window appeared; this informed the user of the current works order number, part number, sample number and component number being examined. A list of all current works order numbers could be displayed if required to assist selection.

At this stage data entry was requested by the system for the sequence of product quality characteristics, the method of data entry (direct from interfaced instruments or keyboard entry) for each being determined by the inspection plan for that part.

When a quality characteristic was inspected the actual value obtained or entered was displayed, together with a visual and audible indication of whether the entered value was acceptable (within tolerance). At the end of each sample inspection operation the user was presented with statistics for the characteristics which had been inspected, and offered the choice of accepting or repeating the inspection, or alternatively of suspending or abandoning the entire inspection sequence. The opportunity to enter a comment was given to the user at the end of each inspection sequence.

The inspection instruction card appeared simultaneously with the inspection record card, and offered instruction as to which dimensions should be inspected, how many readings were required, and the inspection method to be used. Graphics were used to indicate a particular dimension or feature to the user, with a short supporting instruction in text form. The graphics were bit-mapped images obtained either by scanning an existing drawing or by the use of suitable drawing software.

The SPC chart set-up instruction card guided the user through the stages required for setting up the SPC charts. Control limits could be calculated automatically on the basis of process capability from the data entered. It was clear from discussions with potential users that many small batches were inspected which would not justify a chart, and indeed which would not allow a sufficient sample to be obtained to set one up properly. Hence the charts could also be set up using archived data from previous batches of the same part number. This facility had to be used with care, especially as regards process setting. The ability to apply standard or modified control limits on the basis of tolerance information was also provided.

If SPC charts had been set up for a particular part, each subsequent sample would automatically generate new points to be added to the charts and displayed. In the prototype the choice of charts was limited to two popular types for variable data: $\bar{X}$ (sample means) and S (sample standard deviation) Shewhart charts. No attempt was made automatically to assess the existence or nature of any process trend, except that any point displayed outside the control limits would flash to draw the attention of the user.

As regards the inspection plans request card, at the beginning of each weekly scheduling period a new set of inspection instructions could be made available by the CQDB. The CAI workstation would be instructed by the user to acquire the relevant plans via the LAN.

14.4.4 *Advantages of CAI*

Although related to SFDC, to laboratory data acquisition systems and to SPC software, CAI is a coherent concept in its own right. The distinguishing feature of CAI is the combination of direct quality information acquisition, user inspection instructions and the provision of process control information using SPC techniques. A comprehensive CAI application offers a number of advantages in comparison with manual inspection and direct entry methods. These include:

1. reduction of errors caused by misreading of measuring equipment,

entry error, or through deliberate falsification of data;
2. automatic provision of inspection instructions to the user, hence saving time finding the correct specification and minimizing the possibility of using an incorrect version;
3. automatic validation of data on input;
4. immediate automatic comparison with the product specification limits for each variable characteristic;
5. automatic SPC control charting to help detect process trends;
6. comprehensive batch quality reports detailing each characteristic provided automatically, either locally or to a CQDB.

When combined with integration – as a quality data acquisition unit – into an IQS, the CAI workstation becomes not only a tool for the user (process operator or inspector) but a key provider of process and product information to aid the improvement process.

14.5 CONCLUSIONS

This chapter has examined two developing trends in quality data analysis and management: SPC software and SFDC.

SPC assists process control and quality improvement activities by providing a range of techniques for the analysis of quality data. SPC software automates these techniques, cutting out laborious calculations and graph plotting, and making it possible to deal with larger quantities of data. A piece of software is not a replacement, however, for an understanding of the statistical principles involved, or for a comprehensive training programme which places SPC techniques in their proper context as part of a company-wide process of never-ending improvement aimed at TQ.

The latest SPC software is more than just an aid to data analysis for control and improvement. Many recent products are able to store and collate data, so as to provide quality reporting functions. SPC systems are developing towards integrated quality data management, in other words an IQS, but many systems lack a plant-wide approach.

Coming from another direction, SFDC systems are also starting to develop quality data management functions. In the past, most SFDC systems were aimed at quantity data, but the importance of quality has now been recognized. Because they originated as data collection systems for manufacturing functions where process control and improvement are not the central task, SFDC systems tend to lack effective ability in local data analysis. There are some systems, particularly those using portable data collection computers, which are specifically designed for quality data collection.

The final theme of this chapter combines the most useful features of SPC software and SFDC in the form of a general-purpose CAI workstation, which has both local process control and plant-wide reporting functions as an element of an IQS. CAI is an approach mainly used with CMMs, dedicated gauging and ATE equipment, but it may be used with any type of product quality sensor which has the potential to be interfaced to a computer.

As well as providing SPC and reporting functions, the CAI workstation supplies inspection instructions to the user, and automatically evaluates the results of each measurement against product specifications. This is a powerful and flexible combination of abilities which should make the CAI concept of increasing interest to manufacturing industry.

15

Automating quality data management – the quality database

15.1 INTRODUCTION

This chapter examines quality data management in the IQS, and the design and implementation of a quality database. The rationale and background to the choice of a distributed quality system layout, linked to a central database, was explained in Part Two. Here we will discuss practical considerations such as the database contents, hardware and software selection, and computer communications with the quality data collection system.

15.2 EXISTING SYSTEMS FOR QUALITY DATA MANAGEMENT

Most quality data management systems in manufacturing are manual paper-based systems, consisting of a number of independent files. In these files is stored the information required for the quality system, such as:

1. product inspection specifications;
2. inspection sheets for inspection and test results;
3. summaries of product quality information;
4. customer complaints files;
5. calibration files for inspection equipment;
6. receiving goods inspection records;
7. history file of suppliers;
8. non-conformance (scrap, rework and concession) records.

Adherence to quality systems standards requires files such as these, and others, to be maintained. Many quality systems of this type operate adequately, but all too frequently little use is made of these files to provide reports and indicators for decision support. For example, cross-referencing between two paper files to correlate customer complaints with inspection or supplier quality records involves a degree of effort which is only justified when fighting a serious fire!

The introduction of computers into quality data management is being tackled by a small but increasing number of companies, and descriptions of more or less sophisticated data management systems are occasionally found in quality and manufacturing journals. In addition, as explained in the last chapter, some SPC systems include a useful quality data storage and management capability and SFDC systems can also collect and store plant-wide quality data.

Manufacturing and quality assurance staff in many plants have developed data storage collation and reporting systems, usually *ad hoc* and on a small scale, perhaps using spreadsheet software rather than a database. Spreadsheets have been regarded by some users in recent years as an all-purpose computer application, and may frequently be found put to uses for which a relational database would be more appropriate and efficient, especially as regards data correlation and reporting.

15.3 REQUIREMENTS FOR THE QUALITY DATABASE

Obtaining a quality database may be likened to buying a suit. You can buy an off-the-peg version (and perhaps have a bit of minor adjustment done by the vendor), or you may have an application tailored to your requirements, using as a basis some general-purpose database software. With database applications, however, there are more graduations between these extremes, where a variety of levels of individual tailoring are possible. Some systems, although targeted specifically at the quality area, require several man-months of configuration time.

The limitations of the off-the-peg approach are well known to buyers of suits, especially if they are not of a standard size and shape. Somehow the application never seems to fit perfectly, as it was designed with the needs of the general market in mind. The vendor may be able to configure it to your requirements to a certain extent, but serious limitations to flexibility in important areas such as data exchange with other software and computer systems may remain. There are balancing advantages of the off-the-peg quality database, however. You will have a single identifiable supplier who can be expected to maintain your system. In addition, the system will have been initially developed and tested at someone else's expense and inconvenience.

If you choose a more tailor-made route, you will have to be measured to ascertain your requirements. Many are the methods which may be used to define user requirements for a computer database application. If you are fortunate enough to have the services of a competent systems analyst (who also has an in-depth understanding of your company's quality philosophies and systems) to design your application, there should be no cause for concern. Your analyst will design an application to be implemented using a suitable database software package. The choice of database software will be discussed later.

Unfortunately, there is often a gap in understanding between the potential users and the analyst, especially where the application is novel, as is the case in many companies with a quality database. The analyst understands the abilities of the database software, but not the specialist knowledge of the users. The users, with limited computer experience, cannot relate their quality systems knowledge to the potential capabilities of the software. In this situation a prototyping approach can help users define and refine their requirements from the system. By working with an initial prototype, they will gain an improved understanding of the possible, on which they can base requests for changes and enhancements.

Information for the definition of requirements may be obtained from a number of sources, perhaps using the IDEF0 system design process described in Chapter 7. An IDEF0 model can incorporate the perceptions of both systems analyst and potential system users, together with the proposed computer applications, within a context of continuous change and evolution. Lists of data items abstracted from the IDEF0 model may be used as a guide to the data entities required in the quality database.

The use of the IDEF0 technique to define all data flows ensures that database functions are integrated into the overall quality system. Figure 15.1 shows the functions of a quality database. These are:

1. Maintain quality specifications;
2. Receive and correlate batch data;

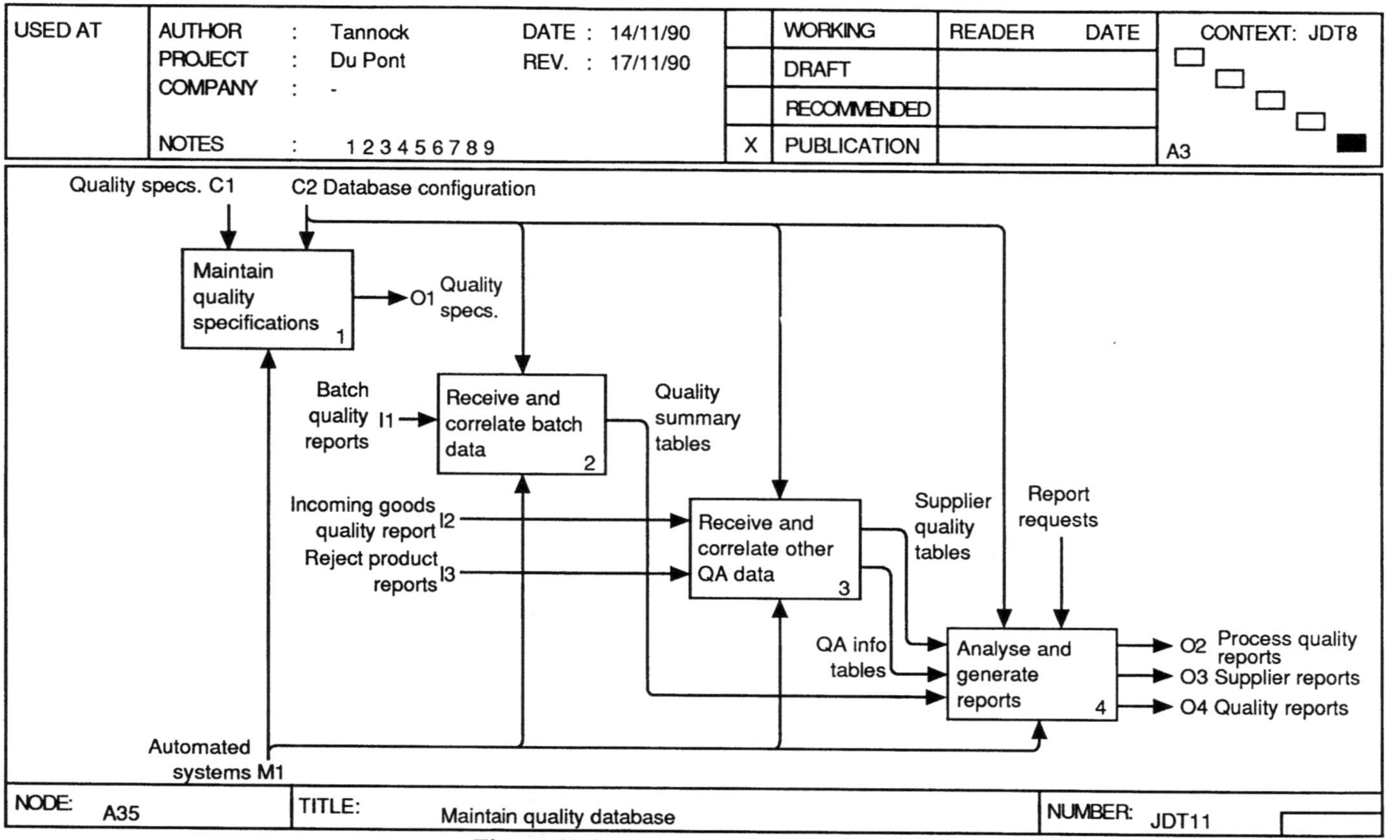

Figure 15.1 Maintain quality database (A35)

3. Receive and correlate other QA data;
4. Analyse and generate reports.

This diagram is a child of 'Perform quality programme' (A3) which appears as Figure 7.4. The inputs to the database are batch quality summaries from manufacturing, incoming goods quality reports and reject product reports. These are correlated into tables available for the analysis and reporting function which is triggered by requests from all who need quality information. As well as storing the quality data from manufacturing, the database also holds the quality plans and specifications against which processes and products are monitored. An important control on all activities is the database configuration.

Our systems analyst might point out at this stage that IDEF0 is not the most suitable way in which to describe and define the structure of a database system. An IDEF0 model can only describe the functions, and does not represent the data entities (such as tables and records) in the database in the most effective way. There are many more appropriate diagrammatic techniques to represent the database design. IDEF1, previously mentioned, is one such technique for describing data entities. As described earlier, IDEF1 is closely associated with IDEF0, and software is available to relate and cross-reference between the two techniques.

If you are tempted to start defining quality database structures without a formal design technique, beware. You will forget something or fail to recognize important relationships between data items.

15.4 FULL AUTOMATION – THE ACTIVE QUALITY DATABASE

A typical database system interacts with the outside world mainly through screen, keyboard and printer. Data are entered by typing on to on-screen 'forms' which can be designed to suit the data formats and arrangements required. The facility to generate data entry forms is provided as part of all database software. The data so entered are placed into tables (locations in the computer memory) by the software, in a way determined by the design of the application. Once stored, they can be correlated with other data and reports produced, either to the screen or to a printer. Hence the level of automation provided by the typical database application is equivalent to that of direct entry inspection or SFDC.

If we wish to use more advanced types of quality data collection, either CAI or full automation, it will be necessary for our database application to communicate in some way with the local computers which carry out the data acquisition tasks. The simplest way to do this is by importing or exporting files containing the required data in the correct format. With most database applications this is only possible by human interaction with the system; the user has to decide which file to import or export and follow a procedure to perform the transaction.

The quality database must be able to import and export quality information to and from the inspection and process monitoring computers which will be required, for example, to report a batch quality summary whenever a batch is finished. Ideally this information should be acquired by the database with the minimum possible manual intervention. This means that certain additional active programs will be necessary, operating on the essentially passive tables of the database.

In multitasking computer systems, programs which perform active functions automatically may coexist and run simultaneously with the database application. They may be triggered by a timer program, by communications received from another process or from an external system. A useful metaphor for understanding the action of the necessary active programs is that of 'servants' to the database function. One approach to the design of the programs required is described below. This approach was conceived and implemented as part of a prototype system developed by the author's research team.

1. The 'butler'. Permanently active, the 'butler' program is designed to wait for the arrival of information, which could be batch summary reports, requests for inspection plans and specifications, or information from other computer systems. When a message arrives, a suitable entry will be placed in a database table (the 'notice board') to alert the 'housekeeper'.
2. The 'housekeeper'. Triggered at regular intervals by a timer function, this process detects the presence of entries placed in the 'notice board' announcing that a batch summary report from a local inspection station has arrived in the file system of the database computer, or that a request for one or more inspection plans had been received. When a notice board entry is detected, the 'housekeeper' will determine the action required by reference to another database table (the 'housekeeper' configuration table)

which holds details of all possible notice board entries and relates each to the name of an active program which should take further action. The appropriate process – known as a 'flunkey' – is then triggered by the 'housekeeper'. In the interests of modularity the functions of the 'housekeeper' were confined entirely to the initiation of the 'flunkey' processes, and she would do no other work herself.

3. The 'flunkeys'. Only triggered by the 'housekeeper', these processes perform specific tasks:
   (a) The 'footman' will carry out the selection of the requested weekly or individual inspection plans and specifications and place the specifications in a file, making them accessible to the remote quality data acquisition stations via network communications.
   (b) The 'porter' will carry out the incorporation of batch quality reports into the various summary tables, from which quality reports are available.
   (c) The 'cleaner' will undertake the removal of entries from the notice board when action has been completed.

Batch quality summary reports will be provided by local inspection and process monitoring computers whenever a batch is completed. The 'butler' process detects the incoming report, recognizes the format, and places an entry in the notice board table. The 'housekeeper' process, when next triggered, reviews the notice board and in turn triggers the 'porter', which takes the batch quality data from the imported file and inserts it into the database summary tables for product quality information. When this task is completed the 'housekeeper' triggers the 'cleaner' process to remove the entry from the notice board.

Using this approach, all that is necessary to implement new functions, such as those allowing functional integration with other manufacturing sub-systems, is the addition of a new 'flunkey' process and the updating of the 'housekeeper' configuration table of the database, via the appropriate database data entry form. Additional active processes could deal with such matters as the incorporation of works order information in files originating from a computerized production management system into database tables, and response to requests from a CAD system for process capability information.

15.5 MAINTENANCE OF INSPECTION PLANS AND SPECIFICATIONS

15.5.1 *Information distribution*

The CQDB will hold all inspection plans and specifications, in addition to receiving summary quality data concerning each batch. But what would happen if a failure of the database or communications network occurred, or if essential maintenance was needed? It would be totally impractical permanently to retain a back-up, paper-based quality data collection system, yet otherwise all inspection would have to cease until the database was back on-line. This potential problem may be alleviated by providing a degree of data distribution on a short-term basis.

The inspection and process monitoring computers will obtain all relevant inspection plans and specifications from the database at the start of each work schedule period (typically one week). Relevant inspection plans are those relating to part numbers referenced by the weekly production schedule file (obtained by the database from the production control system). This arrangement allows the local units to continue working using locally stored specifications. Batch summary reports may be stored until they can be transferred to the database.

15.5.2 *Specification control*

The selection for correct distribution of the inspection plans and specifications is a complex task, especially since it is necessary to take account of constant change in the arrangement of manufacturing and inspection.

The approach adopted must be flexible enough to ensure that automation of quality data collection does not unnecessarily restrict the system users. In many plants, supervisors have considerable autonomy in work scheduling, and can select from the weekly master schedule those jobs they intend to run first. In addition, various scenarios must be considered where batches are unfinished or held for some reason at the end of a scheduling period. These considerations suggest a rule: an automated quality system must not dictate what can be inspected, because this would dictate what is manufactured.

The inspection of a part should be triggered by the user, perhaps the inspector or operator responsible for the process, who wishes to initiate a new batch. By entering a works order identification number from the production schedule, the user will initiate a search for local files to provide an inspection plan. If a plan is not present locally, it must be possible to obtain it by requesting the database to make it available.

The database must be capable of identifying the inspection station type which is to be used to inspect any given stage in the manufacturing sequence. It is necessary to accommodate the usual multistage manufacturing sequence where a single works order number will refer to several stages of manufacture, involving different process machinery and perhaps different inspection station types. All possible work routings must be obtainable from the production control system for this purpose.

An inspection plan identifier code, distinct from the part number, can be allocated to each possible inspection sequence. This will accommodate a situation such as that in which the same part number (possibly from the same batch, and identified by the same works order number) could be made simultaneously on an automatic machine (using automated process monitoring) and on a manual machine (using a CAI workstation for product inspection). These two local inspection systems would require different inspection plans and specifications to be made available from the database.

Attempting to deal with all possible contingencies in the supply of inspection specifications will lead the systems designer to develop a set of interrelated database tables, some of which may be split into header and detail tables, with composite keys for the selection of the inspection details from the appropriate tables. The development of these tables is conceptually complex, making a formal design approach essential.

15.6 QUALITY ASSURANCE INFORMATION

The database can hold information on a range of quality-related matters, in addition to product quality information and inspection plans. Database tables relating to other QA information may be conventional database tables, from which information will be directly entered and accessed using a terminal. Different functions will be selected using a menu screen. A typical menu selection might be:

1. Configuration and maintenance: to support the manufacturing quality database maintenance, database screens must be available to allow review and updating of the configuration tables. A sub-menu will offer the opportunity to view and manually update:
   (a) product and component data;
   (b) works order schedules;
   (c) inspection plans and specifications;
   (d) work routings;
   (e) inspection station allocation;
   (f) inspection station configuration.

2. Direct entry inspection: a sub-menu will offer a selection of direct entry methods for manufacturing quality information. The options include:
   (a) regular inspection;
   (b) special audit inspections;
   (c) receiving goods inspections;
   (d) rework inspections;
   (e) process capability studies.

3. Customer complaints and warranty claims: a sub-menu will offer the following options:
   (a) enter/update customer complaints;
   (b) enter/update warranty claims;
   (c) review complaints awaiting investigation;
   (d) review finalized complaints;
   (e) summary by part number;
   (f) summary by complaint type.

4. Calibration records: inspection instrument calibration records are a requirement of quality systems standards. This option will call up a screen with details of all inspection instruments and their calibration requirements. Selections can be made on the basis of current date to determine which instruments require calibration at any time.

5. Assessment of quality performance: there are various ways to perform this assessment, the most common method being the collection and analysis of quality costs. Chapter 4 examined the various possibilities, and proposed an alternative criterion incorporating a Taguchi loss function. This performance criterion is particularly easy to use with the quality database as it does not

require an extensive cost data collection exercise.

6. Corrective action/improvement tracking: a system for recording the progress of prevention and improvement activities can be very useful. The information recorded may include details of the quality problem which triggered the action, the part numbers affected, customers' or suppliers' involvement, responsibilities and progress against initial time-scales.

7. Other QA records: a number of other useful types of QA records based on the requirements of quality systems standards can be included in the database, such as:
   (a) receiving quality and supplier records;
   (b) audit plans, programmes, and findings;
   (c) material control and traceability records;
   (d) equipment control and maintenance records;
   (e) non-conformance control records;
   (f) process qualification records;
   (g) quality personnel records;
   (h) contract review records.

15.7 GENERATION OF QUALITY REPORTS

The basic manufacturing quality data supplied to the database will allow the generation of a number of reports. Menu selections can offer reporting facilities, such as:

1. quality details of completed batches;
2. non-conformance details;
3. quality summaries for particular periods;
4. quality summaries for particular customers;
5. process capability indices;
6. list of held batches.

Further standard reports can be easily designed and implemented as required. In addition, a knowledge of the database query language will allow a user to select information on the basis of multiple parameters. *Ad hoc* reports could then be provided on, for example, the quality level of all batches for a particular customer during a particular time period.

```
SQL> /
Thu July 20
          Quality Summary for Finished Works Orders
Work         Inspection       Machine        Total    Defective
Order        Spec Used        Number       Defects        Items
---------    ------------     --------     --------   ---------
E1901150     s150             4-1               25           18
             s150             4-3               31           14
                                           --------   ---------
                                                56           32
E1901151     s150             4-1               63           46
                                           --------   ---------
                                                63           46
E1901153     s151             4-2                7            5
                                           --------   ---------
                                                 7            5
E1901154     s100             2-3               36           27
                                           --------   ---------
                                                36           27

SQL  quit
Disconnected from ORACLE
```

Figure 15.2 Quality summary report

Figures 15.2 and 15.3 show two basic text reports generated from a database from the same quality data. Figure 15.2 is the summary showing only the number of defects recorded for four production batches completed on a particular day, while Figure 15.3 shows an expanded report with statistics and capability information relating to essential quality characteristics.

Reports may be presented on-screen, or printed in various text formats. Some database systems have the facility to link with an editor or word processor, which allows high-quality presentation of text to accompany the numerical information. Graphics presentation – for example, of histograms, Pareto analysis or pie charts – is very important to enhance the impact of the information. Some databases do not provide this option, and this should be an important software selection feature.

```
SL> /
Thu Jul 20
         Quality Summary for Finished Works Orders
Work     Machine                         Number
Order    Number  Characteristics Defective      CP    Mean    SD
-------- ------- --------------- ---------    ----   -----  ----
E1901150 4-1     Dimension A          1       1.32    7.75  .025
                 Dimension B          2       1.00    3.05  .033
                 Dimension L          1       2.03   13.34  .016
                 BSPA019 - position   3
                 of end of pin

                 BSPA016 - pins      15
                 parallel

                 Visual straightness  3
         4-3     Dimension A          3       1.05    7.87  .031
                 Dimension B          3       1.76    3.06  .019
                 Dimension L          1       1.97   13.33  .016
                 BSPA019 - position  17
                 of end of pin

                 BSPA016 - pins       5
                 parallel

                 Visual straightness  2
E1901151 4-1     Dimension A          3       1.89    7.72  .017
                 Dimension B         23       3.45    3.06  .009
                 Dimension L         15       2.06   13.32  .016
                 BSPA019 - position  21
                 of end of pin

                 BSPA016 - pins       1
                 parallel
E1901153 4-2     Dimension A          2       1.00   12.83  .050
                 Dimension B          1       2.09    3.05  .024
                 Dimension L          1       2.10   18.42  .015
                 BSPA019 - position   1
                 of end of pin

                 BSPA016 -            1
                 parallelism of pins

                 Visual straightness  1
                 of connector
E1901154 2-3     Dimension A          3       1.89    5.81  .017
                 Dimension L         14       1.45   11.79  .022
                 Pushout Test        15       3.32   26.00  .501
                 Solder Test          4

27 records selected.

SQL> spool out
```

Figure 15.3 Quality capability report

15.8 HARDWARE AND SOFTWARE SELECTION

15.8.1 *Software*

When seeking to purchase a quality database software application, the first place to look for information is the quality and manufacturing press. The various monthly magazines and journals contain articles and advertisements for software which will allow contact to be made with potential vendors. The quality data management systems described will probably be based on general-purpose commercial database software. It is worth knowing something about the pitfalls and potential of such database systems if future problems are to be avoided, and the selection should be made with the benefit of (at least) an outline knowledge of database software itself.

If a system is to be developed from scratch, the basic database software must be chosen from the wide range available. For quality database applications the most suitable type is a relational database management system (RDBMS). This provides greater flexibility in both design and implementation than would be possible with either of the competing network or hierarchical database models.

To ensure maximum flexibility and portability of the finished system, the RDBMS software should be available to run on a wide range of computer hardware and associated operating systems. There may be existing hardware constraining the choice, but this should not lead the buyer to select software which does not come in versions suitable for other computers. Most leading RDBMS software is available to run on a wide range of hardware platforms and operating systems.

Whatever the current hardware, it is important – looking to the future – to choose a product which is available to run under the Unix operating system. Unix is becoming increasingly popular for manufacturing applications because this operating system seems likely to form the basis for a standard, vendor-independent environment for computer applications in manufacturing. The future shape of CIM now looks like a network of Unix workstations and minicomputers sharing the contents of a distributed manufacturing database, of which the quality database will be a vital element.

Unix is a multi-user, multitasking system, which means that a computer using this operating system is able to acquire data from other elements of a distributed system while simultaneously responding to queries addressed to the quality database. A database application developed using a standard microcomputer operating system such as MS-DOS cannot incorporate active programs of the type described earlier, and hence could not achieve fully automatic operation. Microcomputer operating systems which can multitask are available, but are not yet sufficiently advanced to be suitable for use in a distributed system. A version of Unix is available to run on the more capable PC standard computers.

If full automation in data collection is required it will be necessary to develop active 'servant' programs or equivalent functions, as well as the database application. In order to access the database directly with these programs, appropriate functions must be provided with the database software, which must be capable of use with a suitable general-purpose computer language with which active programs can be developed.

Any serious contender for an RDBMS will incorporate features to avoid data loss if transactions with the database fail. This is particularly important when using a distributed system with active elements. A rollback security feature to restore data is essential should a program crash.

Finally, to reduce application development time, the software chosen should provide a comprehensive 4GL program development environment to allow easy prototyping, and also a consistent, well-structured interactive query interface for the system user. The importance of easily generated report graphics has already been mentioned.

15.8.2 *Hardware*

The hardware selection process is constrained by the software requirements described above, and by cost. For most companies the best policy is to start with a small system but ensure expandability. This is particularly important if a prototyping approach is to be used.

Only a few years ago, the computer marketplace was clearly divided into mainframes, minis and micros. Plant-wide database functions demanded a mainframe to meet response time, memory and data storage requirements. To implement an on-line CQDB from scratch for a large manufacturing plant would have required an enormous expenditure. While a mainframe is still an option for the largest plants, most manufacturing companies can provide a quality database much more cheaply with a multi-user workstation.

It has become difficult for 'computer outsiders' to assess the different workstations on offer and compare performance. The functions of minicomputers have in many cases been taken over by workstations, but this term covers a very wide range of performance. Workstations frequently use the Unix operating system, and most are supplied with excellent communications facilities. High capacity disk storage is available.

If a Unix workstation is chosen for the central database, the other elements of the distributed database may be specified as inexpensive micros following the IBM PC standard, and running under the ubiquitous single-user MS-DOS operating system. It is always sensible to specify equipment well known and widely used throughout industry, for which a large amount of software and optional hardware is available. This helps ensure flexibility and economy in configuration and expansion of the system.

15.9 SYSTEM COMMUNICATIONS

Most existing quality database systems use a central minicomputer with a number of standard 'dumb' terminals (without local processing abilities) situated in manufacturing and office areas, connected using serial (RS-232) communications lines. PC-compatible micros can emulate many standard terminals using inexpensive software, so that an automatic inspection or CAI station can contact the database. Unfortunately this type of communication does not usually allow easy and automatic transfer of quality data between programs running on the two computers.

When using a distributed IQS of the type described in Chapter 8, a more effective data communications strategy is essential. The quality database will need to communicate automatically with local computers in various areas of the plant such as incoming goods inspection, manufacturing areas and the desks of quality staff and management.

A LAN is the obvious solution to the problem of linking a number of computers to allow a distributed system to operate efficiently. Any practical LAN in an industrial environment will probably be utilized by other systems for similar tasks. The LAN will require special hardware (in the form of additional boards) and software to be available on each computer. Hardware is often supplied as part of a package with the application software for networking. The actual physical connection will probably be a coaxial or fibre-optic cable.

A number of communication systems exist to link micros and larger computers, and use various underlying protocols to control the interaction between the machines. It is important to choose a system with a standard protocol, which will not inhibit future development. As hardware and software change and develop, the communications system must be capable of expansion to accommodate them. It is desirable that the system chosen is compatible with, or able to be upgraded to, the full Open Systems Interconnection (OSI) standard.

For communications between Unix computers and PC standard micros a number of software products are available. Suitable products allow a micro running under MS-DOS to access disk storage directly on a Unix computer, which acts as the communications 'server'. One such product operates by intercepting MS-DOS system calls, and allowing the user to 'mount' remote file systems, which are physically connected to a server system, but which appear to the MS-DOS micro and its user to be local disk drives. This allows simple and automatic transfer of data files between a micro and the Unix system by MS-DOS programs, such as a CAI workstation application. The communications program remains resident in memory, and takes the appropriate action whenever access to a remote file is requested.

15.10 SUMMARY

Quality data management with a quality database is at the heart of the automated and integrated quality system. Paper-based quality data management systems, although they may be comprehensively designed in accordance with standards, suffer from data correlation, analysis and reporting problems which delay response to problems and limit the information available for decision support and quality improvement. A database can provide a far superior level of quality information.

A database may be implemented using software which varies from dedicated and somewhat inflexible quality database systems, through systems which may be configured to suit various requirements, to those developed from scratch using general-purpose database software. If the last approach is adopted, a complete software development process will be required, and here a quality system design carried out using a structured design technique such as IDEF0 is invaluable.

Most existing quality databases employ direct keyboard entry of quality information to the database, using dumb terminals which display data entry screens. This approach does not allow the integration of automatic and computer-aided local data collection stations. In order to implement an IQS along the lines of that proposed in Chapter 8 these local computers must be able to communicate data automatically to the database computer, and also receive inspection plans and specifications. Dealing with the control and allocation of plans and specifications can be an involved task for a complex manufacturing situation which offers various work routings and alternative processes.

To achieve automatic vertical integration with local inspection and monitoring computers, the database must incorporate active processes to collect and distribute the necessary information. Such processes can be seen as servants to the database. One possible set of servant programs has been described in this chapter.

The contents of the quality database are not limited to product quality data from manufacturing but can include all the information necessary for the quality system – for example, calibration records, customer complaints, and supplier quality records. As well as a variety of reports relating to manufacturing quality performance, further reports can be designed to combine manufacturing quality data with such records, to display linkages and effects.

Hardware and software selection for a quality database should be undertaken with an eye to the future of IT automation in a move towards CIM. Software based on the relational database model is preferable, while hardware able to run the Unix operating system should allow the maximum potential for future developments in integration. Effective computer communications are vital in a distributed system. The use of a LAN with suitable software will allow automatic data transfer between a central database computer and low-cost PC standard micros.

CONCLUSIONS

The automation of quality in manufacturing requires effective systems design firmly underpinned by a convincing strategy for automation. The approach taken must be complementary to a process of TQM.

Effective automation in quality data collection and management is of great value in three critical areas:

1. the operational needs of the manufacturing function, where automation is needed to 'close the loop' rapidly and effectively in a flexible, low-stock manufacturing environment;
2. the information needs of those implementing a process of never-ending improvement as a part of TQM;
3. the integration of manufacturing sub-systems in a move towards CIM, where quality information is one of the crucial business control requirements.

A strategy – the IQS – has been proposed to fulfil the requirements of quality systems automation. The IQS consists of a distributed quality data collection and management system aligned with the information requirements of the business.

This book has also presented a systematic approach to quality systems design, the automation of quality information collection and management, and extensive integration with other automated elements of the manufacturing system.

The field of systems includes theories and techniques which are capable of dealing both with computer developments and with individual and group attitudes and perceptions. A systematic approach should be able to provide a framework for a useful synthesis between disparate themes in manufacturing.

In the course of attempting to shape a strategy for quality systems automation and integration which is compatible with both TQM and CIM, the disparity in aims and means between these two themes in manufacturing thought is plainly exposed. Endeavours to close this split are important to the future success of manufacturing, which must be able to rely on a coherent strategy, rather than a series of changeable fashions. It has been the purpose of this book to make a start in spanning the gap.

APPENDIX A

Derivation of quality performance criteria

This appendix shows the derivation of the quality performance critera for sampling inspection of variable characteristics which was quoted in Chapter 4.

For sampling inspection, where n products are sampled from a batch of size N, we may obtain an estimate of the parameters – the mean (μ) and variance (σ^2) – of a continuous distribution of the deviation (v) from nominal for some variable quality characteristic. Taking into account both the Taguchi loss function and the distribution of this characteristic by means of its probability density function $f(v)$, we can propose that:

$$C = nA + NK \int_{-\infty}^{\infty} v^2 f(v)\mathrm{d}v \tag{A.1}$$

But by the definition of the variance of a continuous distribution represented by a probability density function $f(v)$:

$$\sigma^2 = \int_{-\infty}^{\infty} (v - \mu)^2 f(v)\mathrm{d}v \tag{A.2}$$

By expanding equation (A.2) we obtain:

$$\sigma^2 = \int_{-\infty}^{\infty} v^2 f(v)\mathrm{d}v - 2\mu \int_{-\infty}^{\infty} v f(v)\mathrm{d}v + \mu^2 \int_{-\infty}^{\infty} f(v)\mathrm{d}v \tag{A.3}$$

and rearranging gives:

$$\int_{-\infty}^{\infty} v^2 f(v)dv = \sigma^2 + 2\mu \int_{-\infty}^{\infty} vf(v)dv - \mu^2 \int_{-\infty}^{\infty} f(v)dv \qquad \text{(A.4)}$$

The mean of a continuous distribution is defined as :

$$\mu = \int_{-\infty}^{\infty} vf(v)dv \qquad \text{(A.5)}$$

hence:

$$\int_{-\infty}^{\infty} v^2 f(v)dv = \sigma^2 + 2\mu^2 - \mu^2 \int_{-\infty}^{\infty} f(v)dv \qquad \text{(A.6)}$$

Now since, by definition,

$$\int_{-\infty}^{\infty} f(v)dv = 1$$

we have:

$$\int_{-\infty}^{\infty} v^2 f(v)dv = \mu^2 + \sigma^2 \qquad \text{(A.7)}$$

giving the simple result:

$$C = nA + NK(\mu^2 + \sigma^2) \qquad \text{(A.8)}$$

It should be noted that there is no requirement for the continuous distribution of errors from nominal to be a normal (or any other specified) distribution. It must also be emphasized that cost information of this sort is a comparative quality performance indicator, and must not be confused with the use of variable sampling techniques for acceptance or rejection, although the same data might be used for both purposes.

Because sampling inspection is used, μ and σ^2 in equation (A.8) will only be estimates of the true statistics of the entire batch. The sample mean is the best unbiased estimator of the batch mean. If the sample variance is calculated from measured data using $n - 1$ degrees of freedom for the estimate – that is the denominator in the calculation is $(n - 1)$ – then the sample variance obtained will also be the best unbiased estimator of the batch variance. If specified confidence intervals for the estimates of μ and σ^2 are required, the necessary sample sizes may be easily calculated.

Appendix B
Further reading

Automated inspection

Bretschi, J. (1981) *Automated Inspection Systems for Industry*, IFS Publications, Bedford.

A practical guide to quality data collection from various different types of sensor. Most of the book is still useful and relevant.

Developing special purpose microprocessor systems

Walls, C. (1986) *Programming Dedicated Microprocessors*, Macmillan, London.

A useful introduction to the development of microprocessor firmware aimed at the software engineer, or anyone who is competent in a high-level computer language but needs a guide to low-level programming in the absence of an operating system (Z80 based).

Structured systems design

Ross, D.T. *et al.* (1979) *Architect's Manual – ICAM Definition Method IDEF0*, Softech, Inc., published for ICAM Program Office, Wright Patterson Air Force Base, Ohio.

The definitive and comprehensive practical guide to the use of the IDEF0 technique for structured systems analysis and design. Rather difficult to get hold of at the moment!

CAM-I (1984) *Architecture for Product Assurance*, ICAM Program Office, Wright Patterson Air Force Base, Ohio.

A fully detailed IDEF0 analysis of the (very comprehensive) product assurance system in a US aerospace manufacturing company.

Statistical process control

Owen, M. (1989) *SPC and Continuous Improvement*, IFS Publications, Bedford.

A guide to SPC concepts, techniques and implementation, aimed at never-ending improvement in all areas of the business. Based on many years of experience in SPC training and the introducion of SPC into large and small companies, and illustrated with practical case studies.

Quality management fundamentals

Sinha, M.N. and Willborn, W.O (1988) *The Management of Quality Assurance*, John Wiley and Sons, New York.

One of the better books on quality management. The authors provide a good background of management theory, and appreciate the influence of automation and of other business systems. Also strong on the usual statistical areas.

Appendix C
References

Ackerman, R.B., Plsek, P.E and Surette, G.J. (1986) Quality tools: meeting the needs of tomorrow's industry. *Quality Progress*, October, 61-5.

Alexander, W.H (1983) Commercial success and product quality. *Quality Assurance*, **9(2)**, 31-5.

Bajpai, A.K. and Willey, P.C.T (1989) Questions about quality costs. *Int. Journal of Quality and Reliability Management.* **6(6)**, pp. 9-17.

Bessant, J and Haywood, W. (1986) Integration and advanced manufacturing technology, in *Proc. 4th European Conference on Automated Manufacturing* (ed. B.B. Hundy), IFS Publications, Bedford, pp. 353-64.

Below, L.J. (1987) The meaning of integration, in *Proc. 4th European Conference on Automated Manufacturing* (ed. B.B. Hundy), IFS Publications, Bedford, pp. 345-52.

British Standards Institution (1987; 1990) *BS 5750 Quality Systems*, Parts 0-4, British Standards Institution, London.

British Standards Institution (1990) *BS 6143 Guide to the Economics of Quality*, Part 2, British Standards Institution, London.

British Standards Institution (1987; 1989) *BS 6808 Co-ordinate Measuring Machines,* Parts 1 and 2 (1987), Part 3 (1989), British Standards Institution, London.

CAM-I (1984) *Architecture for Product Assurance*, ICAM Program Office, Wright Patterson Air Force Base, Ohio.

Checkland, P.B. (1984) *Systems Thinking, Systems Practice*, John Wiley and Sons, New York.

Crosby, P.B. (1979) *Quality is Free*, McGraw-Hill, New York.

Crossfield, R.T., Taylor, J., Dale, B.G. and Plunkett, J. (1988) The development of IDEFc as an effective tool for mapping quality management systems, in *Advances in Manufacturing Technolgy III* (ed. B. Worthington), Kogan Page, London.

Cullen, J.M. (1987) Using computers in SPC, in *Proc. International Conference on SPC*, IFS Publications, Bedford, pp. 63-8.

Dessouky, M.I., Kapoor, S.G. and Revor, R.E. (1987) A methodology for integrated quality systems. *Journal of Engineering for Industry*, A.S.M.E. **(109)**, 241-7.

Dodge, H.F. and Romig, H.G. (1941) *Sampling Inspection Tables*, John Wiley and Sons, New York.

Dowding, R. (1987) Automating for quality, in *Proc. International Conference on SPC*, IFS Publications, Bedford, pp. 63-74.

Feigenbaum, A.V. (1983) *Total Quality Control*, 3rd edn, McGraw-Hill, New York.

Flood, R.L. (1988) Review and judgement: measurement and control (systems science) in social contexts. *Trans. Institute of Measurement and Control*, **10(2)**, 79-86.

Godwin A.N., Gleeson J.W.and Gwilliam D.G. (1987) An assessment of the IDEF annotations as descriptive tools. *IMMS Report-8702*, May.

Graham, I. (1988) *Just-In-Time Management of Manufacturing*, Technical Communications, Letchworth.

Hughes, D.R. (1983) *The Development and Application of an Integrated Suite of Models to Aid in the Structure Analysis of Manufacturing Situations*, University of Manchester PhD thesis.

Hughes D.R.and Maull R.S. (1985) A framework for design of CIM system architecture. *Computers in Mechanical Engineering*, September, 34-7.

International Organization for Standardization (1986) *ISO 8402 Quality – Vocabulary*, International Organization for Standardization, Geneva.

International Organization for Standardization (1987) *ISO 9000 Management and Quality Systems Standards – Guidelines for Selection and Use; ISO 9001 Quality Systems – Model for Quality Assurance in Design/Development, Production, Installation and Servicing; ISO 9002 Quality Systems – Model for Quality Assurance in Production and Installation; ISO 9003 Quality Systems – Model for Quality Assurance in Final Inspection and Test; ISO 9004 Quality Management and Quality Systems Elements – Guidelines*; International Organization for Standardization, Geneva.

Jackson, M.C. (1988) An appreciation of Stafford Beer's 'viable systems' viewpoint on managerial practice. *Journal of Management Studies*, **25(6)**, 557-73.

Juran, J.M. (1979) *Quality Control Handbook*, 3rd edn, McGraw-Hill, New York.

Kelly, A. (1989) *Developments at ICL Ashton*, presented at the ACME seminar on Production Management in the Electronics Industry, Aston University, April.

Mahdevieh, Y. (1987) The development of an on-line non-contact visual gauging system, in *Proc. 8th Int. Conf. on Automated Inspection and Product Control* (ed. E. McKee), IFS Publications, Bedford, pp.109-19.

Marsh, J. (1988) *Quality Methods Improvement Analysis*, internal ICL publication.

Maull, R.S. (1986) *An Evaluation of the Contribution of the ICAM Definition Method-IDEF0 to the Analysis and Design of Computer Integrated Manufacturing Systems*, CNAA PhD thesis.

Maull, R.S. (1988) Using the ICAM definition method to model integrated systems of quality control. *Int. Journal of Quality and Reliability Management*, **5(3)**, 29-37.

Mills, D. (1987) Zero defects – a plan to achieve the impossible, in *Proc. 4th European Conf. on Automated Manufacturing* (ed. B.B. Hundy), IFS Publications, Bedford, pp. 235-46.

Newell, D. (1990) *A Study of How Companies Introduce, Develop and Measure the Process of Quality Improvement*, Manchester School of Management MSc thesis, UMIST.

Oliver, N. (1986) Computers and quality, in *Quality Systems* (course PT622), Open University Press, Milton Keynes.

Opdebeeck, E.J. and Oakland, J.S. (1983) Factors in the development of computers for quality control, in *Proc. 22nd Quality Assurance Conference*, IQA, London.

Owen, M. (1987) SPC – education and training implications, in *Proc. International Conference on SPC*, IFS Publications, Bedford.

Owen, M. (1989) *SPC and Continuous Improvement*, IPS Publications, Bedford.

Ranky, P.G. (1987) Real time quality control loops in CIM environment, in *Proc. 8th Int Conf. on Automated Inspection and Product Control* (ed. E. McKee), IFS Publications, Bedford, pp. 45-62.

Rembold, U., Blume, C. and Dillman, R. (1985) *Computer-Integrated Manufacturing Technology and Systems*, Marcell Dekker, New York.

Ross, D.T. (1977) Structured analysis (SA): a language for communicating ideas, *IEEE Transactions on Software Engineering*, **SE-3(1)**,16-34.

Ross, D.T.*et al.* (1979) *Architect's Manual – ICAM Definition Method 'IDEF0'*, Softech, Inc. for ICAM Program Office, Wright Patterson Air Force Base, Ohio.

Rzevski, G. (1987) The concept of CIM system architecture, in *Proc.4th European Conf. on Automated Manufacturing*, (ed. B.B. Hundy), IFS Publications, Bedford, pp. 365-78.

Schneiderman, A.M. (1986) Optimum quality costs and zero defects: are they contradictory concepts? *Quality Progress*, November, 28-31.

Schonberger, R.J. (1986) *World Class Manufacturing – The Lessons of Simplicity Applied*, Free Press, New York.

Shewhart, W.A. (1931) *Economic Control of Quality of Manufactured Product*, Van Nostrand, New York.

Shingo, S. (1986) *Zero Quality Control*, IFS Publications, Bedford.

Sinha, M.N. and Willborn, W.O. (1985) *The Management of Quality Assurance*, John Wiley and Sons, New York.

Stephens, R. and Fox, D. (1987) IDEF modelling application to quality assurance assessments, in *Trans. 41st Annual Quality Congress*, ASQC, Milwaukee.

Taguchi, G. (1981) *On-Line Quality Control during Production*, Japanese Standards Association, Tokyo.

Tuttle, J. (1987) Computer-aided quality, in *Trans. 41st Annual Quality Congress*, ASQC, Milwaukee.

Vail, P.S. (1988) *Computer-intergrated manufacturing*, Delmar Pub. Co., USA.

Walls, C. (1986) *Programming Dedicated Microprocessors*, Macmillan, London.

Willborn, W. (1986) Quality assurance and computer-assisted production. *International Journal of Quality and Reliability Management* **2**(**2**), 18-24.

Withers, L. (1990) Seconds out – a case study of quality management software. *Industrial Computing*, June, 24-27.

Wortberg, J. (1989) Quality assurance with injection moulded products. *PE International – Industrial and Production Engineering*, **13(3)**.

Yeomans, R.W. (1987) Design rules and development guidelines for CIM projects, in *Proc. 4th European Conf. on Automated Manufacturing* (ed. B.B. Hundy), IFS Publications, Bedford, pp. 395-412.

INDEX